Tunisia-Japan Symposium: R&D of Energy and Material Sciences for Sustainable Society

(TJS 2014)

Journal of Physics: Conference Series Volume 596

Gammarth, Tunisia
28 – 30 November 2014

Editors:

K. Akimoto
Y. Suzuki
M. Monirul Islam

ISBN: 978-1-5108-0253-7
ISSN: 1742-6588

Printed from e-media with permission by:

Curran Associates, Inc.
57 Morehouse Lane
Red Hook, NY 12571

Some format issues inherent in the e-media version may also appear in this print version.

Printed by Curran Associates, Inc. (2015)

For permission requests, please contact the Institute of Physics
at the address below.

Institute of Physics
Dirac House, Temple Back
Bristol BS1 6BE UK

Phone: 44 1 17 929 7481
Fax: 44 1 17 920 0979

techtracking@iop.org

Additional copies of this publication are available from:

Curran Associates, Inc.
57 Morehouse Lane
Red Hook, NY 12571 USA
Phone: 845-758-0400
Fax: 845-758-2634
Email: curran@proceedings.com
Web: www.proceedings.com

Tunisia-Japan Symposium: R&D of Energy and Material Sciences for Sustainable Society

(TJS 2014)

Journal of Physics: Conference Series Volume 596

Gammarth, Tunisia
28 – 30 November 2014

ISBN: 978-1-5108-0253-7
ISSN: 1742-6588

IOP Institute of Physics

TABLE OF CONTENTS

TJS2014 Tunisia-Japan Symposium: R & D of Energy and Material Sciences for Sustainable Society

Editorial Preface

This volume of the Journal of Physics: Conference Series contains papers presented at the Tunisia-Japan Symposium: R&D of Energy and Material Sciences for Sustainable Society (TJS 2014) held at Gammarth, Republic of Tunisia on November 28-30, 2014. The TJS 2014 is based on the network of the Tunisia-Japan Symposium on Science, Society and Technology (TJASSST) which has been regularly organized since 2000. The symposium was focused on the technological developments of energy and materials for the realization of sustainable society. To generate technological breakthrough and innovation, it seems to be effective to discuss with various fields of researchers such as solid-state physicists, chemists, surface scientists, process engineers and so on. In this symposium, there were as many as 109 attendees from a wide variety of research fields. The technical session consisted of 106 contributed presentations including 3 plenary talks and 7 key-note talks. We hope the Conference Series and publications like this volume will contribute to the progress in research and development in the field of energy and material sciences for sustainable society and in its turn contribute to the creation of cultural life and peaceful society.

Editors

Katsuhiro Akimoto
University of Tsukuba, Japan

Yoshikazu Suzuki
University of Tsukuba, Japan

M. Monirul Islam
University of Tsukuba, Japan

Journal of Physics: Conference Series **596** (2015) 011001 doi:10.1088/1742-6596/596/1/011001

TJS2014 Tunisia-Japan Symposium: R & D of Energy and Material Sciences for Sustainable Society

Symposium photo:

Journal of Physics: Conference Series **596** (2015) 011002 doi:10.1088/1742-6596/596/1/011002

Peer review statement

All papers published in this volume of *Journal of Physics: Conference Series* have been peer reviewed through processes administered by the proceedings Editors. Reviews were conducted by expert referees to the professional and scientific standards expected of a proceedings journal published by IOP Publishing.

Dye-sensitized solar cells using double-oxide electrodes: a brief review

Yoshikazu Suzuki,[1,2] Yuji Okamoto[2] and Natsumi Ishii[2]

[1]Faculty of Pure and Applied Sciences, University of Tsukuba, 1-1-1 Tennodai, Tsukuba, Ibaraki, 305-8573, Japan.
[2]Graduate School of Pure and Applied Sciences, University of Tsukuba, 1-1-1 Tennodai, Tsukuba, Ibaraki, 305-8573, Japan.

E-mail: suzuki@ims.tsukuba.ac.jp

Abstract. Dye-sensitized solar cells (DSC or DSSC) have been widely investigated because of their potentially high cost performance compared with Si-based solar cells and of their fascinating appearance. DSC with photoelectric conversion efficiency of >10 % (or even 12 %) have been reported, where porous TiO_2 films are generally used as semi-conductor electrodes. Such porous TiO_2 films usually have high specific surface area, and thus, they adsorb plenty of dye molecules, resulting in high photocurrent density. Recently, some double oxides have been examined as alternative photoanode materials, mainly in order to improve photovoltage. Here, studies on DSC using double-oxide electrodes, i.e., perovskite, spinel, ilmenite, wolframite, scheelite and pseudobrookite-types, are briefly reviewed.

1. Introduction

After the pioneer work on dye-sensitized solar cells (DSC or DSSC) by O'Regan and Grätzel in 1991 [1], wide variety of oxide-based semiconductors, instead of TiO_2, have been studied as alternative electrode materials of DSC. Here, studies on DSC using double-oxide electrodes are briefly reviewed with focusing on the simple replacement for TiO_2, *i.e.*, double-oxide-based phase-pure porous electrode (type A in Fig. 1 [2]) .

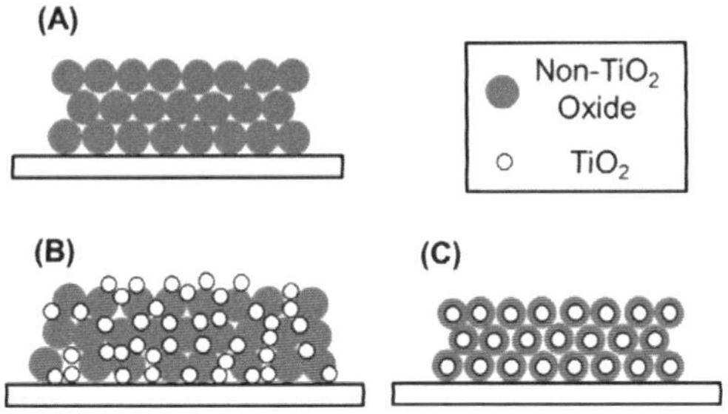

Figure 1. Non-TiO_2 oxides used for DSC electrodes: (A) phase-pure porous electrode, (B) composite electrode with TiO_2, and (C) coating materials on TiO_2. Reproduced with permission from ref. [2] Copyright 2014 The Ceramic Society of Japan.

2. Perovskite-type oxides (ABO_3)

Among various double oxides, perovskite-type oxides have been most widely studied due to the band-structure similarity to TiO_2 anatase. In 1999, Burnside et al. [3] synthesized nanocrystalline mesoporous $SrTiO_3$ by hydrothermal treatment of nanocrystalline TiO_2 and $Sr(OH)_2$ at 190°C for 4.5 h, and applied the obtained nanocrystalline $SrTiO_3$ to photoanodes of DSC by sintering at 500-600°C in a flowing oxygen atmosphere. They confirmed that the dye uptake of the $SrTiO_3$ films was about half of the TiO_2 anatase film, which suggests that the surface treatment of $SrTiO_3$ or the optimization of dye may improve the cell performance.

In 2001, Lenzmann et al. [4] reported $SrTiO_3$ nanocrystalline photoanode, which was evaluated by surface photovoltage spectroscopy. They found a blue shift for the $SrTiO_3$-based cell as compared to the TiO_2 anatase based cell. In 2010, Yang et al. [5] determined the flat band edges (E_{fb}) of nanostructured $SrTiO_3$ electrodes in propionitrile (PN), acetylacetone (Acac) and PN/Acac mixture. They also prepared N3-dye sensitized $SrTiO_3$ and evaluated its photovoltage-photocurrent curves (Table 1). They reported short-circuit current density (J_{SC}), open-circuit voltage (V_{OC}) and *I-V* curves of N3-dye sensitized $SrTiO_3$, but the power conversion efficiency (PCE) was not stated in detail.

In 2014 (online in Oct. 2013), based on the Yang's work [5], Jayabal et al. [6] pointed out that (1) both TiO_2 and $SrTiO_3$ have similar band gap of value of ~3.2 eV, but the flat-band potential of $SrTiO_3$ is greater than that of TiO_2 anatase, and (2) the conduction band of $SrTiO_3$ lie ~0.2 eV above the conduction band of TiO_2 anatase, hence $SrTiO_3$ is expected to produce excellent photovoltage. To demonstrate the feasibility of $SrTiO_3$, Jayabal et al. synthesized mesoporous $SrTiO_3$ via hydrothermal treatment of $Sr(CH_3COO)_2$ and $Ti[OCH(CH_3)_2]_4$ at 120°C for 12 h with post calcination at 900°C for 6 h, and they applied it to DSC (Table 1). High V_{OC} of 0.73 eV is ascribed to the more negative positioned conduction band of $SrTiO_3$, and low current density is attributed to the reverse saturation current.

Also in 2014, Okamoto et al. [2] reported the synthesis of $SrTiO_3$, $CaTiO_3$ and $BaTiO_3$ submicron powders by solid-state reaction at 1200°C for 2 h followed by planetary ball-milling, and they applied the powders to DSC (Table 1). Without the optimization of redox electrolyte etc., PCE values of this research were much lower than other works. Still, they succeeded to demonstrate the difference among three perovskite-type electrodes. Note that the $SrTiO_3$ and $BaTiO_3$ powders were effective as second phases of composite type electrodes (type-B in Fig. 1). In 2015 (online in Oct. 2014), Li et al. [7] synthesized Ho^{3+}-doped $SrTiO_3$ in order to obtain the down-conversion effect and p-type doping effect. Although PCE values of their study were also much lower than other works, apparent increase in J_{SC} was confirmed. Besides titanates, other perovskite-type oxides have been also studied, such as barium stannate ($BaSnO_3$) with band gap of 3.1 eV[8,9]. Shin et al. reported the PCE of 4.5% for bare $BaSnO_3$.

Table 1. DSC using perovskite-type double oxide electrodes.

Electrode	Electrode area (cm^2)	Electrode thickness (μm)	Type of dye	Light source (mW/cm^2)	V_{OC} (V)	J_{SC} (mA/cm^2)	FF	PCE η (%)	References
$SrTiO_3$	–	5-6	N3	~100	0.789	3	0.70	1.8	Burnside (1999) [3]
$SrTiO_3$	0.196	–	N3	100	0.6	0.41	–	–	Yang (2010) [5]
$SrTiO_3$	1	–	EY*	62.5	0.73	0.44	0.55	0.51(?)	Jayabal (2014) [6]
$SrTiO_3$	1	~ 10	N719	100	0.636	0.128	0.464	0.038	Okamoto (2014) [2]
$CaTiO_3$	1	~ 10	N719	100	0.488	0.0172	0.312	0.0026	Okamoto (2014) [2]
$BaTiO_3$	1	~ 10	N719	100	0.650	0.127	0.433	0.036	Okamoto (2014) [2]
$SrTiO_3$	0.25	–	N719	100	0.54	0.15	0.51	0.04	Li (2015) [7]
Ho^{3+}-doped $SrTiO_3$	0.25	–	N719	100	0.60	0.23	0.49	0.07	Li (2015) [7]
$BaSnO_3$ (CP)	0.4	~ 10	N719	100	0.68	2.77	0.60	1.1	Guo (2010) [8]
$BaSnO_3$ (HT)	0.4	~ 10	N719	100	0.67	2.04	0.59	0.8	Guo (2010) [8]
$BaSnO_3$ (SSR)	0.4	~ 10	N719	100	0.58	0.37	0.47	0.1	Guo (2010) [8]
$BaSnO_3$	0.16	–	N719	100	0.61	11.19	0.66	4.5	Shin (2013) [9]

*EY: Eosin Yellow, CP: coprecipitation, HT: hydrothermal, SSR: solid-state reaction.

Table 2. DSC using spinel, ilmenite, wolframite, scheelite and pseudobrookite-type double oxide electrodes.

Electrode	Electrode area (cm^2)	Electrode thickness (μm)	Type of dye	Light source (mW/cm^2)	V_{OC} (V)	J_{SC} (mA/cm^2)	FF	PCE η (%)	References
Spinel									
Zn_2SnO_4	0.20	5.6	N719	~100	~0.62	~9.1	~0.67	3.8	Tan (2007) [10]
Zn_2SnO_4	0.32	5.7	N719	100	0.68	2.4	0.68	1.0	Lana-Villarreal (2007) [11]
Mg_2TiO_4	1.0	30.8	N719	100	0.586	0.033	0.293	0.0056	Ishii (2015) [12]
Ilmenite									
$CdSnO_3$	0.2	15	N719	~100	0.43	7.43	0.45	1.42	Natu (2010) [13]
$MgTiO_3$	1.0	28.8	N719	100	0.515	0.048	0.317	0.0079	Ishii (2015) [12]
Wolframite									
$ZnWO_4$	0.25*	~12*	N3*	100	0.57	0.21	0.459	0.054	Kim (2011) [14]
$MgWO_4$	0.25*	~12*	N3*	100	0.46	0.18	0.429	0.036	Kim (2011) [14]
Scheelite									
$CaWO_4$	0.25*	~12*	N3*	100	0.41	0.13	0.422	0.023	Kim (2011) [14]
$SrWO_4$	0.25*	~12*	N3*	100	0.39	0.12	0.433	0.020	Kim (2011) [14]
Pseudobrookite									
$MgTi_2O_5$	1.0	42	N719	100	0.465	0.060	0.248	0.0069	Ishii (2015) [12]

*Jung et al. [15]

3. Spinel-type oxides (AB_2O_4)

Spinel-type double oxide group (general formula: AB_2O_4) also contains many compounds. Some spinel-type oxides have been examined as photoanodes of DSC. In 2007, Tan et al. [10] examined zinc stannate, Zn_2SnO_4 (inverse spinel type, $Zn[Zn,Sn]O_4$) with band gap of 3.6 eV, and reported the PCE of 3.8%. According to Tan et al., the electron diffusion length for Zn_2SnO_4 is shorter than P25 TiO_2 film. However, for thin films ($\leq \sim 6$ μm), the photocurrent density for a Zn_2SnO_4 cell was higher than that for a P25 cell with the same film thickness. Lana-Villarreal et al. [11] also reported the DSC performance of Zn_2SnO_4 DSC.

In 2015, Ishii et al. [12] reported Mg_2TiO_4 (inverse spinel type) with band gap of 3.7 eV as a photoanode of DSC. In their preliminary data, since Mg_2TiO_4 particles were large and film thickness was not optimized (too thick as a photoelectrode), DSC performance was rather low. Similar to Zn_2SnO_4, the electron diffusion length for Mg_2TiO_4 may be shorter than TiO_2. Further optimization will be expected.

4. Ilmenite-type oxides (ABO_3)

When the ionic radii of A and B ions in ABO_3 are similar to each other, ilmenite-type double oxides preferably form instead of perovskite-type. Some ilmenite-type oxides also have been studied as photoanodes of DSC. In 2010, Natu and Wu [13] studied cadmium stannate, $CdSnO_3$ with band gap of 3.07 eV, and reported the PCE of 1.42%. The conduction band edge of $CdSnO_3$ is ca. 0.9 eV lower than that of TiO_2 anatase. A merit of $CdSnO_3$ double oxide is its higher stability in an acidic environment than SnO_2. Ishii et al. [12] also preliminary reported on $MgTiO_3$ with band gap of 3.7 eV as a photoanode of DSC.

5. Wolfamite and scheelite-type oxides (MWO_4)

In 2011, Kim et al. [14,15] prepared divalent metal tungstates, MWO_4, with wolframite (M=Zn and Mg) and scheelite (M=Ca and Sr) structures by a solid-state reaction method, and applied them for DSC. Wolframite-type showed better performance than scheelite-type, which is attributable to their enhanced electron transfer resulting from appropriate band positions. The band gaps of $ZnWO_4$, $MgWO_4$, $CaWO_4$ and $SrWO_4$ are 2.95, 3.48, 4.25 and 4.58 eV, respectively. Correspondingly, the PCE of $ZnWO_4$, $MgWO_4$, $CaWO_4$ and $SrWO_4$ were 0.054, 0.036, 0.023 and 0.020%, respectively.

6. Pseudobrookite-type oxides (M_3O_5)

Another double oxide in MgO-TiO_2 binary system, $MgTi_2O_5$ (pseudobrookite, band gap of 3.4 eV) was also preliminary studied by Ishii et al. [12].

7. Conclusion

Studies on DSC using perovskite, spinel, ilmenite, wolframite, scheelite and pseudobrookite-type double oxides are briefly reviewed. Common positive aspect using double oxide will be their flexibility of band tuning. Perovskite-type $BaSnO_3$ and spinel-type Zn_2SnO_4 are most promising to date. Compared with conventional TiO_2-based DSC, however, further optimization is required.

Acknowledgment

This work was supported by Grant-in-Aid for Science Research No. 23350111 (Basic Research: Category B), University of Tsukuba-NIMS Collaborative Research Project, and JST/ACT-C "Creation of methanol synthesis catalysts working at room temperature based on mechanistic principles of CO_2 activation" head by Prof. Junji Nakamura.

References

[1] O'Regan B and Grätzel M 1991 *Nature* **353** 737.
[2] Okamoto Y and Suzuki Y 2014 *J. Ceram. Soc. Jpn.* **122** 728.
[3] Burnside S, Moser J E, Brooks K, Grätzel M and Cahen D, 1999 *J. Phys. Chem. B* **103** 9328.
[4] Lenzmann F, Krueger J, Burnside S, Brooks K, Grätzel M, Gal D, Ruhle S and Cahen D 2001 *J. Phys. Chem. B* **105** 6347.
[5] Yang S M, Kou H, Wang J C, Xue H B and Han H L 2010 *J. Phys. Chem. C* **114** 4245.
[6] Jayabal P, Sasirekha V, Mayandi J, Jeganathan K and Ramakrishnan V 2014 *J. Alloy Compd.* **586** 456.
[7] Li Y Y, Hao H S, Qin L, Wang H L, Nie M Q, Hu Z Q, Gao W Y and Liu G S 2015 *J. Alloy Compd.* **622** 1.
[8] Guo F A, Li G Q and Zhang W F 2010 *Int. J. Photoenergy* **2010** 105878.
[9] Shin S S, Kim J S, Suk J H, Lee K D, Kim D W, Park J H, Cho I S, Hong K S and Kim J Y 2013 *ACS Nano* **7** 1027.
[10] Tan B, Toman E, Li Y G and Wu Y Y 2007 *J. Am. Chem. Soc.* **129** 4162.
[11] Lana-Villarreal T, Boschloo G and Hagfeldt A 2007 *J. Phys. Chem. C* **111** 5549.
[12] Ishii N, Okamoto Y and Suzuki Y 2015 *Int. Lett. Chem. Phys. Astro.* **7** 9.
[13] Natu G and Wu Y Y 2010 *J. Phys. Chem. C* **114** 6802.
[14] Kim D W, Cho I S, Shin S S, Lee S, Noh T H, Kim D H, Jung H S and Hong K S 2011 *J. Solid State Chem.* **184** 2103.
[15] Jung H S, Lee J K, Lee S, Hong K S and Shin H 2008 *J. Phys. Chem. C* **112** 8476.

Enhanced light trapping using plasmonic nanoparticles

L Manai[1], B Dridi Rezgui[1], A Bou[2], D Barakel[2], P Torchio[2], R Benabderrahmane Zaghouani[1], O Palais[2] and B Bessais[1]
[1] Photovoltaic Laboratory, Research and Technology Center of Energy, Borj Cedria Science and Technology Park 2050 Hammam-Lif, Tunisia
[2] Institut Matériaux Microélectronique Nanosciences de Provence-IM2NP, Aix Marseille Université, CNRS-UMR 7334, Domaine Universitaire de Saint Jérôme, Service 231, 13397 Marseille Cedex, France

E-mail: bechir.rezgui@crten.rnrt.tn

Abstract. Plasmonics is a new light trapping method used in photovoltaic (PV) solar cells. A significant enhancement of the scattered and absorbed incident light due to the use of silver nanoparticles (Ag-NPs) was observed, which yield to the exaltation of the electromagnetic field in the vicinity of these NPs. In this context, we investigate optically and morphologically the effect of the NPs size dependence on the localized surface plasmon resonance. Extinction, absorption and scattering cross sections are calculated using Mie theory.

1. Introduction

Plasmonic field has recently attracted attention in silicon-based photovoltaic devices due to its exceptional properties [1-4]. Interacting with sun light, metal NPs can support localized surface plasmon resonance (LSPR) caused by the collective oscillations of their conduction electrons. The excitation of surface plasmon leads to an extraordinary scattering [5] and a strong enhancement of local field in the vicinity of the NPs surface [6, 7]. This enhancement increases the optical path of light inside the active layer [8]. This considerably contributes to the increase of the photo-generated current as more light is absorbed and collected. Noble metals like gold (Au) and silver (Ag) are best candidates to support localized surface plasmon, where light can resonantly be scattered or absorbed in accordance with NPs size, shape, inter-distance and surrounding medium [9-11]. Silver would be a better choice than gold because it has the higher scattering efficiency as well as the lower wavelength interband transition [12, 13].

Modeling the optical response of Ag-NPs interacting with sun light can be based on Mie theory [14]. NPs are supposed to be spherical (radius R) in a non absorbing medium (air with a refractive index n_m). Within this theory, the extinction σ_{ext} , scattering σ_{sca} and absorption σ_{abs} cross sections are expressed in series expansion of the involved fields [15], which are described in terms of spherical harmonic functions, such that, the different multi-polar excitations and their contribution can be easily identified:

$$\sigma_{ext} = \frac{2\pi}{k^2} \sum_{j=1}^{\infty} (2j+1)\,\mathrm{Re}(a_j + b_j) \tag{1}$$

$$\sigma_{sca} = \frac{2\pi}{k^2}\sum_{j=1}^{\infty}(2j+1)(|a_j|^2 + |b_j|^2) \tag{2}$$

$$\sigma_{abs} = \sigma_{ext} - \sigma_{sca} \tag{3}$$

Where $k = (2\pi N)/\lambda$, λ is the wavelength of the incident light, N is the refractive index of the medium surrounding the particle, a_j and b_j are defined as follows:

$$a_j = \frac{m\psi_j(w)\psi_j'(v) - \psi_j(v)\psi_j'(w)}{m\psi_j(w)\xi_j'(v) - \xi_j(v)\psi_j'(w)} \tag{4}$$

$$b_j = \frac{\psi_j(w)\psi_j'(v) - m\psi_j(v)\psi_j'(w)}{\psi_j(w)\xi_j'(v) - m\xi_j(v)\psi_j'(w)} \tag{5}$$

In these expressions, ψ_j and ξ_j are Ricatti-Bessel functions and dimensionless parameters are defined as:

$$m = N / n_m \tag{6}$$

$$v = k / R \tag{7}$$

$$w = kc / n_m \tag{8}$$

In this work, we investigate in a first part the effect of Nps size on the enhancement of scattered or absorbed light. Then we study experimentally the effect of mass thickness on the optical and morphological properties of Ag-Nps.

2. Experimental details

Silver (Ag) thin films were deposited on silicon substrates (p-type, (100) oriented, 300 μm-thick) by thermal evaporation technique. In order to eliminate organic and inorganic impurities from the surface, silicon wafer was ultra-sonicated sequentially in deionized water, acetone and ethanol for 15 mn at 40 °C. Ag with mass thicknesses of nearly 5, 10, 15 and 20 nm was then evaporated at chamber pressure of 10^6 mbar while maintaining a low deposition rate (0.1 Å/s) in order to cover homogenously the silicon surface. To break continuous film into agglomerates, samples were annealed under argon atmosphere during 60 min at 250°C. In order to investigate the effect of NPs size on the light scattering, absorption, scattering and extinction cross sections of various radiuses are calculated using Mie Theory (matlab code). Total reflectance was measured using a Perkin Elmer Lambda 950 UV-Vis-NIR (250 – 2500 nm) spectrophotometer equipped with an integrating sphere. JEOL FEG Scanning Electron Microscope (SEM) was used to study the morphology of Ag-NPs. Images taken from SEM are analyzed quantitatively and qualitatively using an image processing and analysis software (*Image J*, Ref[16]). Our study investigates the effect of particles size, film thickness and coverage rate on the optical properties including extinction, reflection and LSPR peaks shift related to metal Nps.

3. Results and discussions

Focusing on the change in the NPs size, LSPR peaks shift was theoretically investigated using Mie theory [17]. Incident light interact effectively with Ag-NPs for cross sections larger than the geometrical cross section [18]. In fact, the size of the NPs has a considerable effect on LSPR and consequently on their optical properties. Two different mechanisms can be distinguished inducing size dependence of the LSPR properties. Below 50 nm of radius, light tends to be absorbed than scattered [19]. For this size range, we may assume that NPs are properly described by a dielectric dipole. Size dependence of the LSPR mainly affects the intensity of the resonance band while the effect on the resonance wavelength is quite reduced leading to a small shift towards lower frequencies (figure 1a).

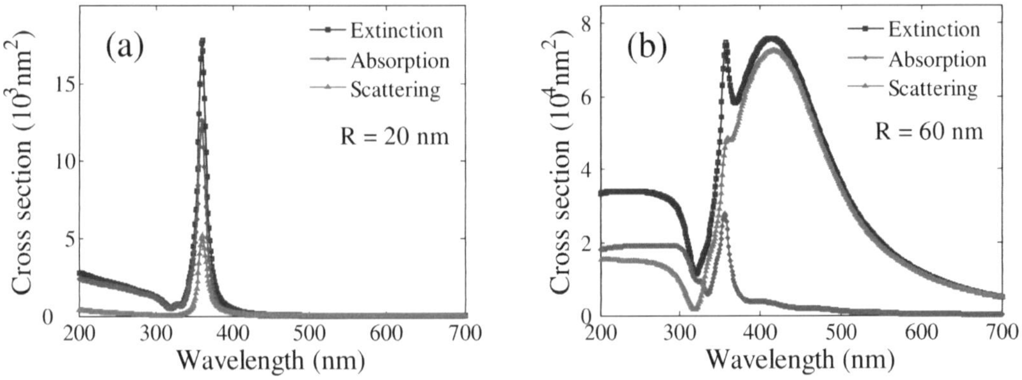

Figure 1. Calculated extinction, absorption and scattering cross sections of Ag-NPs in vacuum (air) with 20 nm (a) and 60 nm (b) radius. Measured values for the dielectric function $\varepsilon(\omega)$ are taken from Johnson and Christy database [18].

Up to 50 nm, light wavelength is quite larger than NPs size. Therefore, Nps are not well described by a dipole and then multi-polar terms are required. For larger particle sizes, the excitation of higher-order plasmon modes must be taken into consideration. The spectral response is modified due to both retardation effects and appearance of high order mode in the Mie scattering coefficients (figure 1b). The resonance band splits into several bands: one band for dipolar, two bands for quadrupolar, three bands for octopole, etc. In this case, scattering pre-dominates absorption since retardation effects become more and more important. Figure 2 depicts the normalized extinction cross section for different radius of Ag-NPs. We can clearly observe the red-shift and broadening of the resonance band with the increase of the particle size. This makes the Ag-NPs a very efficient tool for solar cells as we can manipulate light trapping and scatter most of the radiation with the largest angular spread in order to maximize the optical path length.

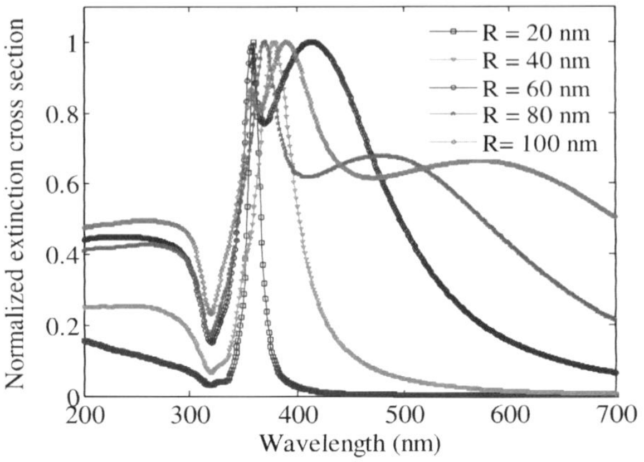

Figure 2. Calculated extinction cross section of Ag-NPs in vacuum (air) with 20, 40, 60, 80 and 100 nm radius.

According to Mie theory calculation of absorption and scattering cross section, it is clear that the spectral resonance position and strength of the dipole as well as of the high-order mode depend noticeably on the dimension of the NPs. To correlate with experiment, a morphological study is necessary. For this purpose, we deposited Ag thin films of nearly 5, 10, 15 and 20 nm thicknesses on silicon substrate. Annealing steps under a neutral atmosphere (argon) is necessary to break continuous films into agglomerated islands. Annealing was carried out at 250°C during 60 min. Shape, size and surface coverage rate of Ag-NPs depend on the thickness of the deposited films and annealing parameters (time and temperature). Table 1 shows morphological characteristics derived from SEM images and analyzed by *Image J* software.

Table 6. Morphological characteristics of samples with different mass thicknesses.

Mass thickness (nm)	**5**	**10**	**15**	**20**
Surface coverage rate (%)	4	9	57	74
Average particles size (nm)	18	25	60	80

Figure 3 shows SEM images (a, b, c and d) and size distribution (e,f,g and h) of Ag-Nps. By increasing the mass thickness, and then the NPs size, the surface coverage rate significantly increases from 4 % to 74 %. Furthermore, Ag-NPs become more interconnected, elongated and larger in size. The average particles size increases from 18 nm to 80 nm for 5 and 20 nm mass thickness respectively. It is obvious from figure 3d (20 nm mass thickness) that the annealing parameters are not sufficient enough to compress the NPs and to transform elongated NPs to regular spheroids.

The optical response of Ag-NPs depends essentially on size, shape and surface coverage rate. Figure 4 shows the total reflectance for Ag-NPs layers deposited on silicon substrate with various mass thicknesses. For 5 nm thickness, the resonance modes are lost due to the weak coverage rate. NPs are slightly distributed on the surface of silicon, as can be seen in figure 3a. The reflectance spectrum corresponding to 5 nm mass thickness has almost the same aspect than that of the silicon substrate. While for 10 nm mass thickness, which corresponds to NPs size of about 25 nm, the first peak located at 365 nm appears. According to Mie theory, this can confirm the dependence of the resonance mode on NPs size for dimensions below 50 nm. For the two other samples (15 nm and 20 nm mass thickness), two resonance peaks appear due to the excitation of the high order mode. The shorter wavelength peaks maximum corresponds to quadrupolar resonance whereas longer wavelength maximum peak corresponds to dipolar resonance. The peak of dipolar related band is red-shifted and broadened from 500 nm to 580 nm as the mass thickness increase from 15 nm to 20 nm. The red-shift leads to an increase of the reflectance intensity of the maximum of the dipolar peak, which in turn is due to the retardation effects of the electromagnetic field across the nanoparticles as their size increases [21].

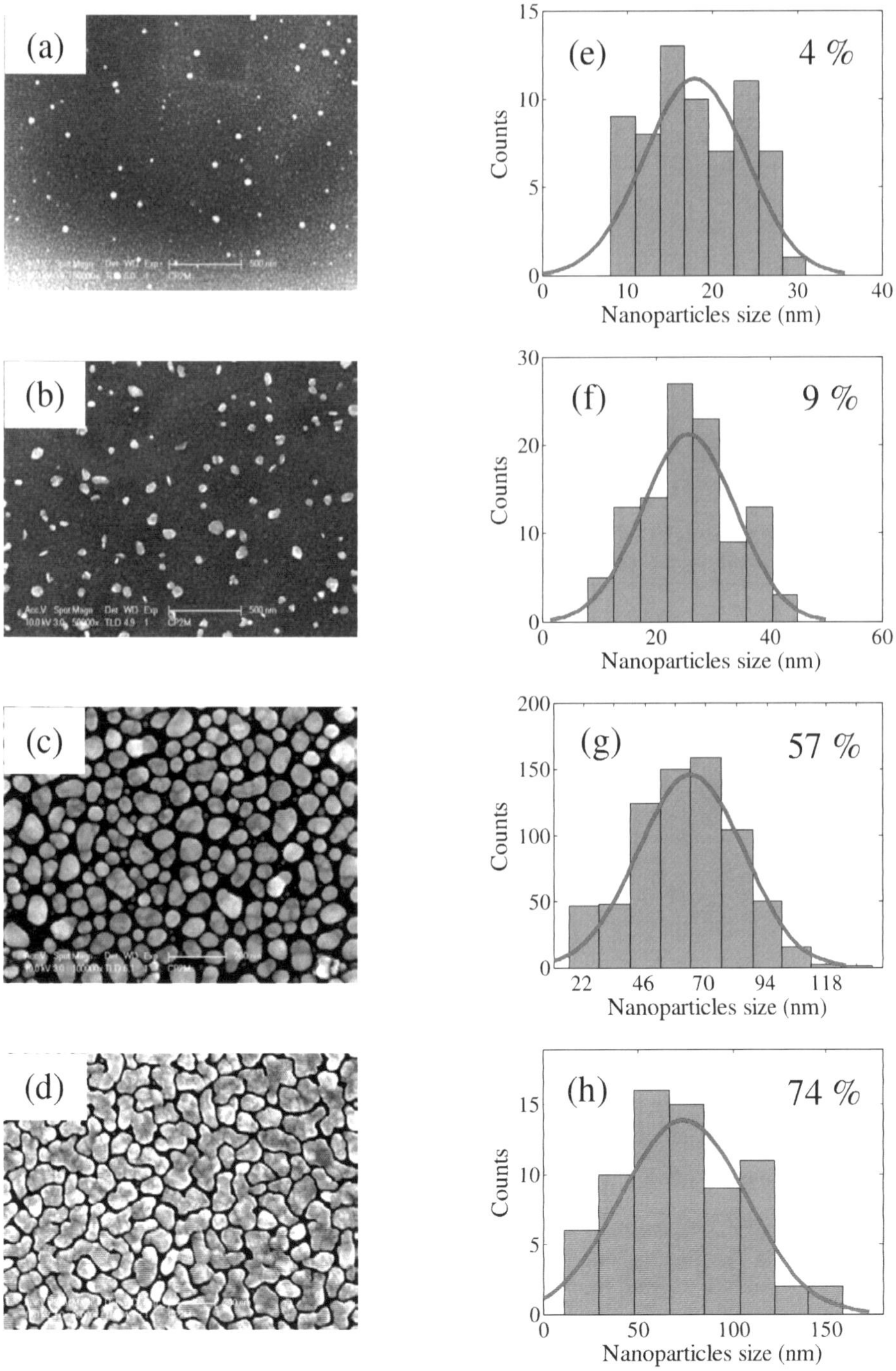

Figure 3. SEM images of Ag-NPs with mass thicknesses corresponding to (a) 5 nm, (b) 10 nm, (c) 15 nm and (d) 20 nm grown on silicon substrates and annealed at 250° C during 60 min. (e), (f), (g) and (h) correspond to the NPs size distribution, respectively.

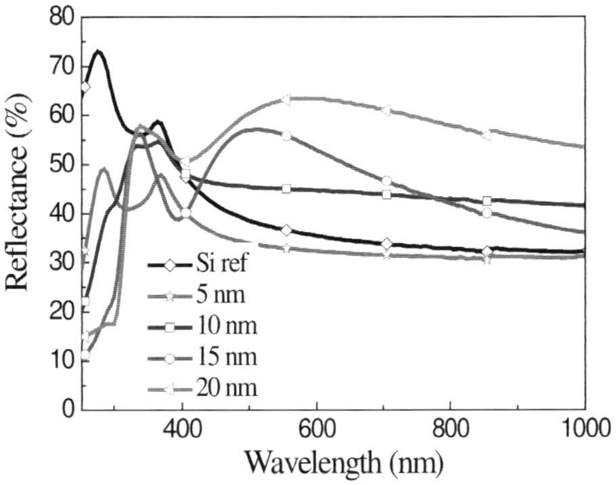

Figure 4. Measured total reflectance spectra of Ag-NPs layers with various mass thicknesses deposited on silicon substrates.

4. Conclusion

The resonant nature of plasmonic enhancement of the Ag-NPs is investigated in this work. We theoretically demonstrate that the absorbed part of light by small NPs is much higher than the scattered part, while scattering is much important for larger Nps due to the excitation of high order modes. With the increase of mass thickness of silver thin films deposited by thermal evaporation and depending on the size of annealed Ag-Nps, experimental reflectance spectra are proven to be mainly affected.

5. References

[1] Catchpole K R and Polman A 2008 *Opt. Express.* **16** 21793
[2] Atwater A and Polman A 2010 *Nat. Mat.* **9** 205
[3] Adamovic N and Schmidt U *Electrotechnik & Informationstechnik* **128** 342
[4] Duche D, Torchio P, Escoubas F, Monestier F, Simon J J Flory F and Mathian G 2009 *Sol. Energ. Mat. Sol. Cells* **93** 1377
[5] Luk'yanchuk B S, Tribelsky M I, Wang Z B, Zhou Y, Hong M H, Shi L P and Chong T C 2007 *Appl. Phys. A* **89** 259
[6] Kottmann J P, Martin O J F, Smith D R and Schultz S 2000 *Opt. Express* **6** 213
[7] Ewe W B, Chu H S and Li E P 2007 *Opt. Express* **15** 18200
[8] Catchpole K P and Polman A 2008 *Appl. Phys. Lett.* **93** 191113
[9] Kelly K L, Coronado E, Zhao L L and Schatz G C 2003 *J. Phys. Chem. B* **107** 668
[10] Noguez C 2007 *J. Phys. Chem. C* **111** 3806
[11] Vedraine S, Torchio D, Duche D, Flory F, Simon J J, Le Rouzo J and Escoubas L 2011 *Sol. Energ. Mat. Sol. Cells* **95** 57
[12] Murray W A and Barnes W L 2007 *Adv. Mater.* **19** 3711
[13] West P R, Ishii S, Naik G V, Emani N K, Shalaev V M and Boltasseva A 2010 *Laser Photon. Rev.* **4** 795
[14] Mie G 1908 *Ann. Phys. (Leipzig)* **25** 377
[15] Bohren C F and Huffman 1998 *Absorption and Scattering of Light by Small Particles* (New York: Wiley-Interscience)

[16] http://rsb.info.nih.gov/ij/index.html
[17] Fu Q. And Sun W 2001 *Appl. Optics* **40** 1354
[18] Akimov Y A, Ostrikov K and Li E P 2009 *Plasmonics* **4** 107
[19] Santbergen R, Temple T, Liang R, Smets A, Swaaij R and Zeman M 2012 *J. Opt.* **14** 024010
[20] Johnson P B and Christy R W 1972 *Phys. Rev. B* **6** 12
[21] Thouti E, Chander N, Dutta V and Komarala V K 2013 *J. Opt.* **15** 035005

Water reservoir as resource of raw material for ceramic industry

M Irie[1] and J Tarhouni[2]

[1]Alliance for Research on North Africa (ARENA), University of Tsukuba, 1-1-1Tennodai, Tsukuba, Ibaraki, Japan.
[2]National Institute of Agronomy, 43, Avenue Charles Nicolle 1082 Mahrajène, Tunis, Tunisia.

E-mail: irie.mitsuteru.fu@u.tsukuba.ac.jp

Abstract. The industries related to the ceramics such as construction bricks, pottery and tile are the important sectors that cover the large part of the working population in Tunisia. The raw materials, clay or silt are excavated from opencast site of limestone clay stratum. The opencast site give the negative impact on landscape and environment, risks of landslide, soil erosion etc. On the other hand, a most serious problem in water resource management, especially in arid land such as Tunisia, is sedimentation in reservoirs. Sediment accumulation in the reservoirs reduces the water storage capacity. The authors proposed the exploitation of the sediment as raw material for the ceramics industries in the previous studies because the sediment in Tunisia is fine silt. In this study, the potential of the water reservoirs in Tunisia as the resource of the raw material for the ceramics industries is estimated from the sedimentation ratio in the water reservoirs.

1. Introduction

The ceramic industries are key industries for Tunisia that has only limited resources in all sectors generally. The boom in building construction recent years expanded the demand of construction bricks, tiles and other construction materials made of ceramics. Seeing the tourism sector that is the most important sector in this country, crockery is major souvenir, especially produced in Nabeul and Sejnane. The sector of construction material and ceramics in Tunisia made the biggest growth rate (3.2%) after Chemical industry (15.3%), while the textile industry and electric machine industries are shrinking due to the European debt crisis. However, the resource acquisition for ceramics influences the local environment. The excavation sites of the clay are opencast cutting thorough a hill (Figure 1). These works lose the landscape and increase the risk of soil erosion.

On the other hand, from the point of view of water resource management, one of the important issues is management of water reservoirs. The capacity loss of surface water resources by sedimentation in North African countries reaches 0.5% of total storage capacity in Morocco, 0.5% in Algeria, and 1.0% in Tunisia per year respectively, estimated approximately [1]. In the most serious case, almost or all of the storage capacity of the dam was occupied by sediment (Figure 2). The authors proposed the exploitation of the sediment in the reservoirs as the raw material of construction bricks in the previous study [2]. We performed the venting test with the test pieces of the ceramics made of the sediment samples from the reservoirs. And then, we tried to evaluate the water resource recovery effect.

However, we went no further than the comparison between the capacity of the dredging machine and clay demand in the factories for the balance estimation of supply - demand. The total volume of the sediment that accumulate on all the reservoirs in Tunisia were not estimated. In this study, the annual yield of sediment from each reservoirs in Tunisia is estimated from the geographical and hydrological information. Finally, it is compared with the demand of the clay resource in Tunisia.

Figure 1. Opencast site of the clay

Figure 2. El Kebir dam occupied by sediment

2. Material and methods

Ben Mammou and Louati [3] estimated the volume of the sedimentation of the 19 large dams in Tunisia, based on the bathymetric surveys carried out in different years. However, the estimated annual sedimentation ratio in a reservoir varies between the different terms randomly. That was due to the accuracy of the bathymetric surveys and problem of mapping procedure of the bathymetry. When the initial capacities of the reservoirs were calculated from the ground surface map with elevation contour lines. The volume is calculated by the summation of horizontal thin layers which is figured by the altitude contour line. On the other hand, the capacity volume of a reservoir in use is estimated by the bathymetric survey in cross sections. Then the volume of the portions sandwiched between the cross sections are calculated with the area of each cross section and distance between the sections. This difference of the calculations made the big error.

Thanks for the development of the new apparatuses for the bathymetric survey, nowadays we can obtain more correct data. In the previous study [2], the authors calculate the average annual sedimentation ratio from the result of bathymetric surveys in the 4 reservoirs. In order to estimate the sedimentation ratio of the other 14 reservoirs correctly, of course, the best way is the campaign of the bathymetric surveys with the new apparatus in these reservoirs. It might be carried out in the future, but tentatively, the multi regression analysis was tried. The locations of the reservoirs are shown in Figure 3.

Sedimentation ratio can be approximately estimated from the geographical and hydrological data of the catchment area of the upper rivers of the reservoirs. In fact, Universal Soil Loss Equation (USLE) that is the most famous method for the estimation of soil loss estimation includes rainfall erosivity factor, the soil erodibility factor, the topographic factors and the cropping management factors. Instead of those, catchment area and specific discharge were used as parameters in this study. Those indicate topographic factor and rainfall erosivity respectively.

$$Y = a_1 X_1 + a_2 X_2 + C \qquad (1)$$

Where Y: Annual sedimentation ratio (million m^3/year), X1: Catchment area (km^2), X2: Specific discharge (mm/year).

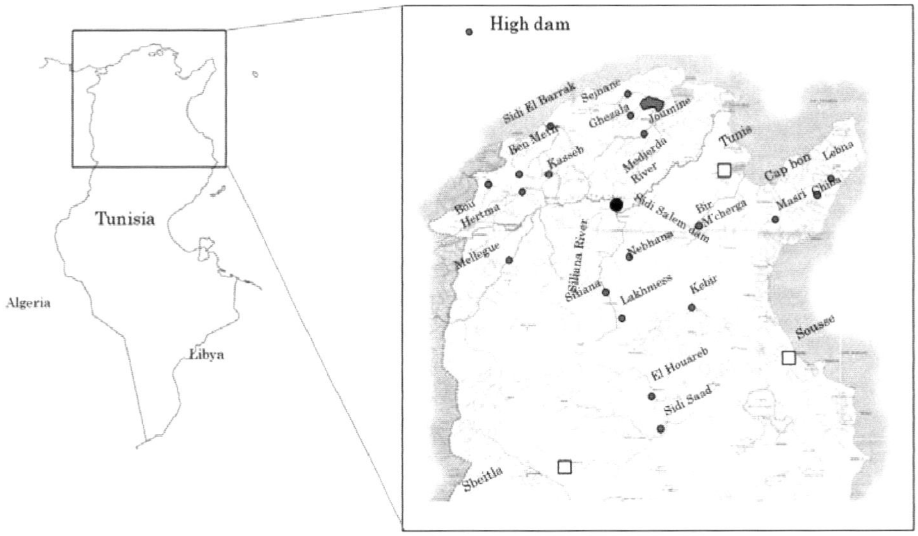

Figure 3. Location of the reservoirs in Tunisia

3. Result and discussion

The characteristics of the sediment samples of the 4 reservoirs have already been reported [2] & [4]. The sediments in the reservoirs are clay and silt with low concentration of organic matter and high concentration of Calcium carbonate. The ceramics test pieces produced from the sediments showed good performance in the vending test for the evaluation of the strength as ceramics. Therefore, the sediment in the reservoirs in Tunisia can be the resource for ceramics industry qualitatively.

The discussion is quantitative balance between annual sedimentation yield and clay demand in ceramics industry. The coefficients of the multi regression equation calculated from the parameters of the 4reservoirs were a_1=3.62*10^{-4}, a_2=7.24*10^{-4}, C=0.078. The sedimentation ratio of the other 14 reservoirs were estimated with this empirical equation. Total volume of the annual sediment yield with 19 reservoirs was 15.6 million m^3.

We assumed that the bulk density of the sediment is 0.65 ton/m^3 averagely, then the weight of it is 10 million ton. One construction brick factory needs 0.22 million ton/year and there are 20 brick factories [2], totally, only 4.4 million ton/year can be consumed and 5.6 million of sediment will be accumulated in the reservoirs.

The ceramics industries which consume the clay is not only brick production, but also pottery or crockery etc. However, it is said that the demand of clay in the pottery which is kind of handcrafts is quite lower than that in the brick production though the pottery is very active as one of the touristic sector in Tunisia. Therefore, the demand of clay in brick production was only counted in this study.

From the point of the view of the ceramics industries in Tunisia, there is plenty supply of raw material, but from that of water resource management, the sustainability of the surface water resource in Tunisia cannot escape from the critical situation even if the exploitation of the sediment for ceramics industries in reservoir is promoted. Needless to say, under the condition that the sediment inflow and accumulation in a year in the reservoirs is larger than the demand of clay, the capacity volume recovery by the excavation of the sediment accumulated from long time ago cannot be expected.

However, the environmental impact of the open pit of the clay excavation site can be reduced by the exploitation of the reservoir sediment.

Table 1 Estimated annual sedimentation Ratio in each reservoir

Name of Reservoir	Catchment area(km^2)	Specific discharge (mm/year)	Sedimentation (M m^3)
Ben Metir	103	427	0.42
Mellegue	10300	17	3.82
Bezirk	75	53	0.14
Chiba	64	109	0.18
Nebhana	855	35	0.41
Lakhmess	127	94	0.19
Masri	40	50	0.13
Kasseb	101	495	0.47
B.M'cherga	1263	35	0.56
Bou Heurtma	390	187	0.35
Sidi Saad	8950	11	3.33
Sidi Salem	7950	56	3.00
Joumine	418	325	0.46
Ghezala	48	292	0.31
Lebna	189	53	0.18
Siliana	1040	56	0.49
El Houareb	1120	38	0.51
Sejnane	367	270	0.41
El Kebir	250	50	0.20
Total			15.6

4. Conclusions

It has already been confirmed that the sediment in the reservoirs are available for the ceramics industries qualitatively in the previous study. In this study, the sediment yield in all the big reservoirs in Tunisia was estimated from the morphological and hydrological information of the reservoirs. The accuracy of the estimation which is calculated with the limited data is controversial. However, even if the error of the estimation is considered, the sediment yield in the reservoirs is obviously larger than the demand of clay in the ceramics industries. Such situation is the sufficient resource supply to the ceramics industries but the sustainability of the surface water resource is critical.

The research group including the authors propose the other exploitation of the sediment for ceramics but water treatment material that is more value added products than construction bricks. Such use of the sediment might make better income which can cover the cost of dredging. However, the oversupply of the sediment might not be solved by the introduction of high value products. Finally, the excess amount of the sediment which cannot be consumed in the ceramics industries can be consumed in the other sectors which have low profitability, such as using as soil amendment [5].

Acknowledgements

This research was partially supported by the Ministry of Education, Science, Sports and Culture, Grant-in-Aid for Scientific Research (A), 2013, 25257306.

References

[1] Remini B 2006 *La Sedimentation Des Barrages* (Universite, Saad Dahlab-Blida)
[2] Irie M, Kashiwagi K, Tarhouni J et al 2012 *J. of Jpn. Soc. of Civ. Eng.:* Series G **68** 6 pp II41-46
[3] Ben Mammou A and Louati M 2007 *Rev. des Sci. de l'Eau* **20** 2 pp 201-211
[4] Irie M et al 2014 *Waste and Biomass Valo.* **5** pp 273-281
[5] Mtibaa S, Irie M, et al 2012 *J. of Arid Land Studies* **22** 1 pp 315-318

Performance evaluation of a solar ejector-vapour compression cycle for cooling application

K Megdouli[1], M Elakhdar[2], E Nahdi[2], L Kairouani[2] and A Mhimid[1]

[1]Laboratoire d'Études des Systèmes Thermiques et Énergétiques Ecole National d'ingénieur de Monastir Avenue Ibn El Jazzar, 5019 Monastir, Tunisie.

[2]Unité de Recherche Energétique et Environnent, Ecole National d'ingénieur de Tunis 37 Le Belvédère Tunis, Tunisie.

E-mail: karima.megdouli@gmail.com

Abstract. This study deals with the performance of the ejector-vapour compression cycle assisted by solar. The effect of operating conditions on the combined cycle performance is examined. Also, a comparison of the system performance with environment friendly refrigerants (R134a, R600, R123, R141b, R142b, R152a, R290, and R245fa) is made. This performance is calculated using an empirical correlation. Thermodynamic properties of functioning fluids are obtained with package REFPROP 8. Using the typical meteorological year file containing the weather data of the city of Tunis, the system performance is computed for three collector types. The theoretical results show that the R290 offers the highest coefficient of performance, COP=3.75, for generator temperature T_B = 78°C, condenser temperature T_c = 30°C and the intercooler temperature T_e = 15°C.

Keywords: Solar energy, Ejector, Compression system, Performance, Combined refrigeration cycle

1. Introduction

An ejector system is relatively simple to construct, to operate and to control. This makes it appears like an attractive solution in the countries having good sunshine. Several research groups have investigated the performance of a simple stage ejector cooling cycle.

Zeren et al. [1] considered the commercial potential of the solar-driven ejector cycle with R12 as the refrigerant. They concluded that the efficiency of such a system depends mainly on the availability of a free solar heat source. Huang et al. [2] have examined the solar ejector cooling system using R141b as the working fluid and have determined the COP~ 0.22 at practical operating conditions for air conditioning. Nahdi et al. [3] studied a solar ejector system for the air-conditioning application, with several environment-friendly working fluids. The result shows that the coefficient of performance of 0.28 can be achieved on an average when the evaporating temperature is at 10 °C.

[1] Corresponding author. Tel.: +216 22507488
E-mail address: karima.megdouli@gmail.com (K. Megdouli).

The COP of the simple stage ejector cooling cycle is still relatively low as a compared to that of vapour compression and absorption systems. The aim of this work is to focus the study on a combined ejector-vapour compression cooling cycle assisted by solar energy that appears as a remarkable alternative solution to improve the COP of the whole system and very little work has been done in this area [4-6]. Sun [7] conducted the mathematical simulation of an environmental friendly solar system. Steam and R134a were used as the refrigerant in an ejector and a compression sub-cycles, respectively. The simulated results showed that the COP of the system could be improved up to 50% compared to the conventional vapour compression system. The system consisted of a solar collector, hot water storage tank, a conventional compression and an ejector sub-cycles with heat exchanger as an interface between them. The COP drops dramatically and even to zero when it is under atrocious weather. In order to prevent the failure of solar system, an auxiliary burner, driven by gas or oil, may be added to ensure that the pressure and temperature conditions required by the ejector are guaranteed. The water carries heat from the collector and releases it in the storage tank. Then, the hot water is pumped from the storage tank to the generator which vaporizes the refrigerant.

In case the heat delivered from the storage tank is not enough to drive the cycle, additional heat is produced in the auxiliary burner and delivered to the R290. At the same time, the vapour leaving the intercooler enters into the ejector and generates a mixed stream that is discharged into the condenser. This saturated liquid is divided into two streams; one goes into the pump where is pumped back to the generator and the other goes to the expansion valve. It is then expanded to the intercooler where it is evaporated by the heat rejected from the vapour mechanical compression cycle.
At the mechanical subsystem, the compressed R290 vapour coming from the compressor is condensed in the intercooler, then exposed to pressure reduction in the throttling valve and then enters the evaporator where it is evaporated to produce the necessary cooling effect. The vapour is finally compressed to a higher pressure by the compressor and then enters to the intercooler, thus completing the combined cycle. A solar energy assisted combined ejector-vapour compression system is schematically shown in Fig.1.

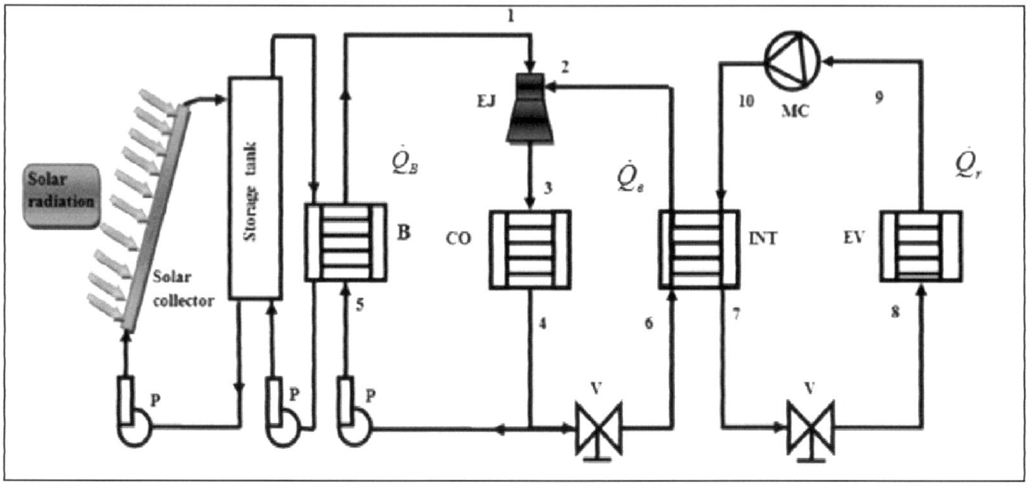

Figure 3. Schematic of a solar ejector-vapour compression system

2. Solar collector performance

The efficiency of a solar system depends on the collector type, solar radiation intensity and the system operating conditions. The solar collector efficiency η_c is defined as the ratio of the gain of the useful heat over any time period to the solar radiation over the same period. It was defined by Duffie and Beckmain [8] as:

$$\eta_c = F_R(\tau\alpha) - F_R U_L \frac{T_i - T_a}{I} \quad (1)$$

Where $F_R(\tau\alpha)$ and $F_R U_L$ are the average heat removal factor and the heat loss coefficient (W m^{-2}K^{-1}), respectively. They change depending on the collector type.

T_a : The ambient temperature.

I : The total solar radiation on a flat surface per unit area.

T_i : The collector's inlet temperature.

In simulation models found in the literature T_i is considered to be 10°C higher than the generator temperature [2].

Pridasawas and Lundqvist [9] proposed three solar collectors for the heat supply of the solar ejector system to investigate the effect on the system's performance and comparison.

Type 1 is a single glazed flat plate solar collector:

$$\eta_c = 0.8 - 5\frac{T_i - T_a}{I} \quad (2)$$

Type 2 is a double glazed flat plate solar collector:

$$\eta_c = 0.8 - 3.5\frac{T_i - T_a}{I} \quad (3)$$

Type 3 is an evacuated-tube solar collector:

$$\eta_c = 0.8 - 1.5\frac{T_i - T_a}{I} \quad (4)$$

3. Modeling of the ejector :

The ejector is the heart of the ejector refrigeration sub-cycle. This component replaces the compressor in the conventional compression refrigeration cycle. The ejector consists of four parts: a nozzle section for primary flow, a section chamber for the secondary flow, a mixing section for the primary and secondary flows, a throat section and the diffuser section for the mixed fluid to recompress to the previous pressure Fig.2.

The refrigerant vapour enters into a primary nozzle where the energy of pressure is converted into velocity energy in a suitable experimental condition; the driving pressure ratio corresponds to the supersonic flow of the jet in the divergent part of the primary nozzle. The high speed jet creates the entertainment force for entrainment of secondary fluid then occurs and the velocity energy of the mixture is converted into pressure in the secondary tube. The mixture at higher pressure returns to the condenser.

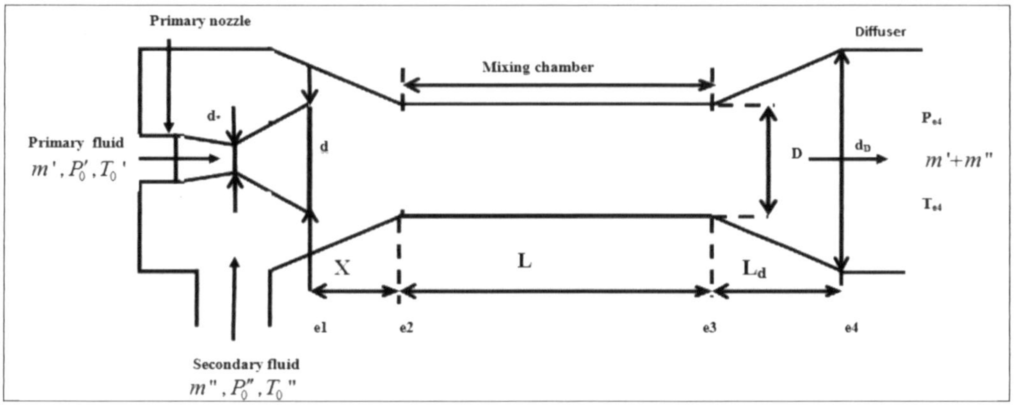

Figure 3. Schematic view of an ejector

Nahdi et al. [10] Holton and Schiltz [11], Work and Haedrich [12] shown that a great number of parameters that determine the performance of the ejector are related to the nature of the fluid, the thermodynamical properties and the ejector geometry.
Lu [13] showed that the basic thermodynamical parameters are the entrainment ratio (U) (the flow rate of the entrained vapour divided by the flow rate of the motive vapour). The expansion ratio (Γ), it is defined as the ratio of the motive vapour pressure to the entrained vapour pressure. The compression ratio (r) gives the pressure ratio of the compressed vapour to the entrained vapour. The driving pressure ratio (ξ) is defined as the ratio of the motive vapour pressure to the back pressure, and the main geometrical parameter ratio Φ (ratio of the areas of the mixing chamber to the primary nozzle throat).

On the basis of the previous experimental investigation, an empirical correlation was derived for the performance prediction of ejectors in optimal regime and independent of the fluid nature [14].

$$U = U_{opt} = 3.32\left[\frac{1}{r} - \frac{1.21}{r\xi}\right]^{2.12} \tag{5}$$

4. Performance analysis of a combined ejector-vapor compression cycle

In the configuration shown in Fig. 1, the solar heat that drives the refrigeration system is determined by the operation temperature T_i. The combined cycle consists of two sub-cycles. The heat exchanger which is an evaporator of the ejector sub-cycle is used to increase the COP of the overall system. Its working principle is depicted in [4-15].
Let's define the COP_s of the combined cycle, the ejector cycle and the vapour compression cycle as:

$$COP = \frac{\dot{Q}_r}{\dot{Q}_B + \dot{W}_{comp}} \tag{6}$$

$$COP_{ej} = \frac{\dot{Q}_e}{\dot{Q}_B} \tag{7}$$

$$COP_m = \frac{\dot{Q}_r}{\dot{W}_{comp}} \tag{8}$$

The overall thermal efficiency of the ejector sub-cycle is defined as the ratio of the refrigeration $\dot{Q}_e$ to the solar radiation on the collectors [3].

$$COP_s = \frac{\dot{Q}_e}{\dot{Q}_s} = \frac{\dot{Q}_e.\dot{Q}_B}{\dot{Q}_B.\dot{Q}_s} = \eta_c COP_{ej} \quad (9)$$

However, it should be noticed that the efficiency of the solar collector efficiency η_c defined as $\dot{Q}_B / \dot{Q}_s$ will depend on the heat transfer process which takes place with phase-change in the generator heat exchanger and therefore will also depend on the outlet refrigerant temperature, T_B.

5. Results and discussion

In the present work, the computer program is developed and simulated with FORTRAN. The refrigerant properties are evaluated by using REFPROP V8.0 [16]. This study used $8 \leq T_e \leq 26$ °C and $30 \leq T_c \leq 42$°C. The isentropic efficiencies of the pump and the compressor are assumed to be 0.85 and 0.6 respectively.

5.1. *Effect of fluid nature on the COP*

Some properties of the selected refrigerants are given in table 1. Comparative calculations have been made for R134a, R600, R123, R141b, R142b, R152a, R290, and R245fa.

Table 1: some properties of the selected working refrigerants

Name	NBP	ODP relative to R11	GWP relative to CO_2
R134a	-26.1	0	1600
R600	-0.6	0	<10
R123	27.81	1.0	77
R141b	32.05	0.1	630
R142b	-9	0.043	2300
R152a	-24.3	0	140
R290	-42.2	0	3
R245fa	14.9	0	820

Fig.3 represents the COP of the solar refrigeration system for various refrigerants. It can be seen that the performance varies from refrigerant to refrigerant. The refrigerant R290 gives the highest COP (COP=3.75) for the operating temperatures (T_B = 78°C, T_e = 15°C, T_c=30°C).

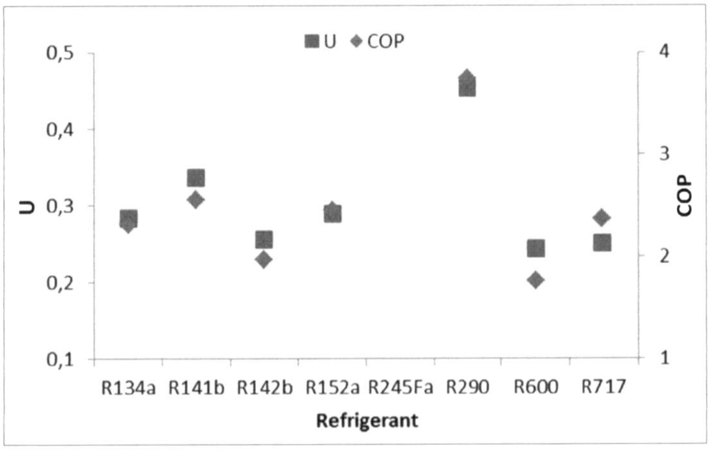

Figure 3. The entrainment ratio U and the COP of combined ejector cycle for various refrigerants

5.2. *Effect of T_e*

The variations of the performance of ejector cycle COP_{ej} and the combined cycle performance COP with T_e are represented in Fig.5.

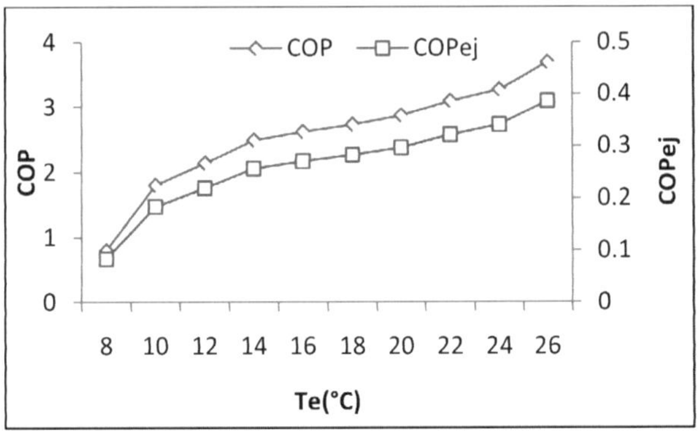

Figure 5. Coefficient of performance of the combined system against T_e

The ejector entrainment ratio, COP_{ej} and COP increase with T_e. Increasing T_e leads to an increase in the secondary flow on the one hand and to a reduction in the latent heat of evaporation on the otherhand. The increasing ratio in the secondary flow is higher and therefore, the COP and the COP_{ej} increases with an increasing T_e for fixed boiler and condenser temperature.

5.3. *Effect of T_c*

The condensing temperature, T_c, mainly depends on the environmental temperature. The following results were obtained by varying the condensing temperature from 30 °C to 42 °C with T_e=10 °C and

T_B=78 °C. The effect of the condenser temperature on the COP is examined. As indicated in Fig.6, the increase in the condenser temperature causes the COP to decrease. Since the same amount of vapor cannot be compressed to a higher condenser pressure, COP will decrease.

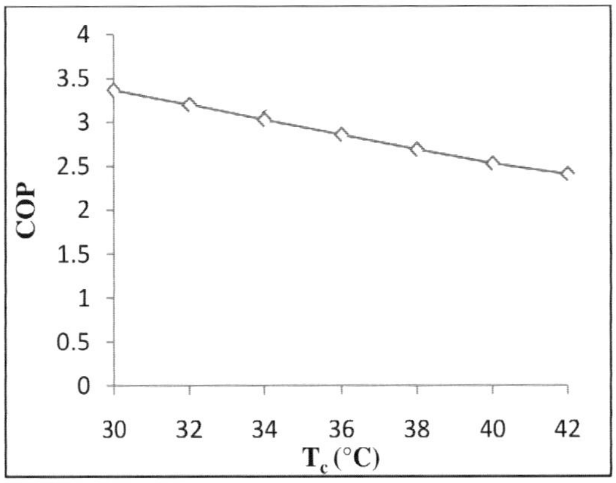

Figure 6. Effect of T_c on the COP

5.4. *Effect of I:*

The meteorological data were supplied by the Tunisian Institute of Meteorology for the city of Tunis [17]. These data considered of total solar radiations on horizontal surface for a representative day. The incidence angle of the collector is taken as 30°. For all months, the hourly solar radiation intensity incident upon the collector surface was calculated from the meteorological data [17] and presented in Fig. 7. Maximum solar radiation takes place as I=1030 W m^{-2} at 12.00 in August.

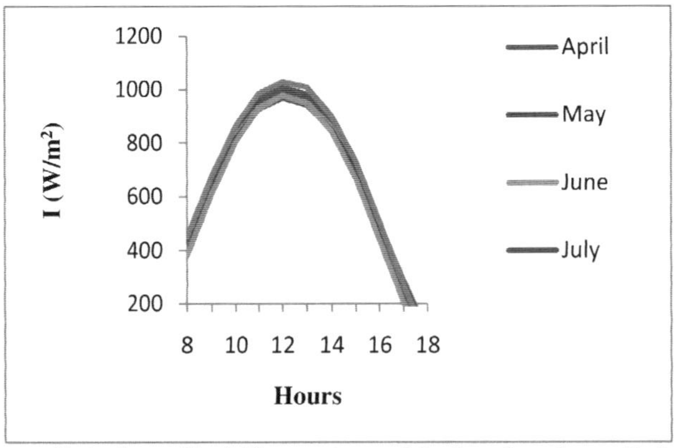

Figure 7: (color online) Comparison of the hourly solar radiation

The monthly average of the efficiency of the solar collector is given in Fig.8. The highest monthly efficiency of the flat plate collector η_c is determined as 0.7 in August.

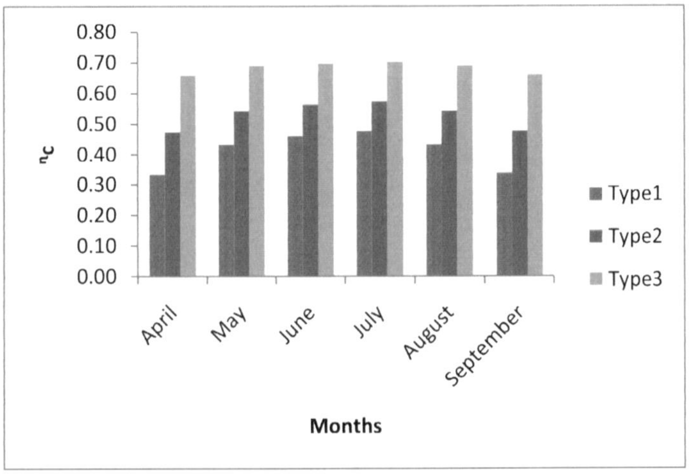

Figure 8: Monthly average efficiency of three flat plate collectors

The hourly variations of the collector efficiency are shown in Fig. 9. As seen, the collector efficiency depends on the collector type. It increase and reaches maximum value at 12:00 with the rise of the solar radiation. The evacuated-tube solar collector provides higher efficiency.

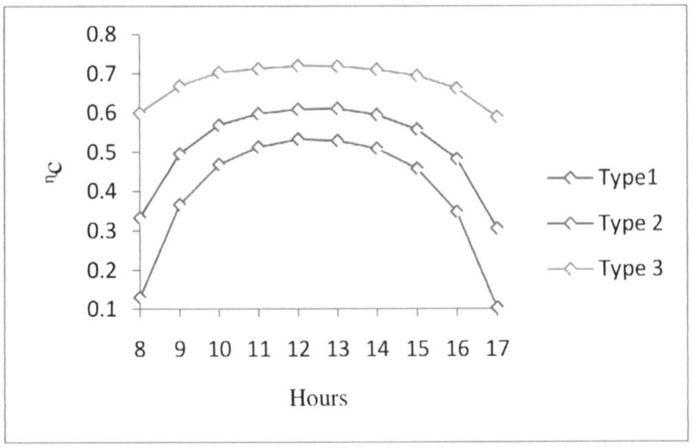

Figure 9: Hourly variations of efficiency of three flat plate collectors

Figures 10 and 11 show the hourly and monthly performances of the solar combined ejector system when the system operates with different types of solar collectors. The evacuated-tube solar collector gives higher COP_s values than a single glazed flat plate solar collector and a double glazed flat plate solar collector for the operating temperatures T_B=78 °C, T_e=10 °C and T_c=35 °C. The highest COP_s is determined to be 1.9 for collector type1, 2.3 for collector type 2 and 2.8 for collector type 3.

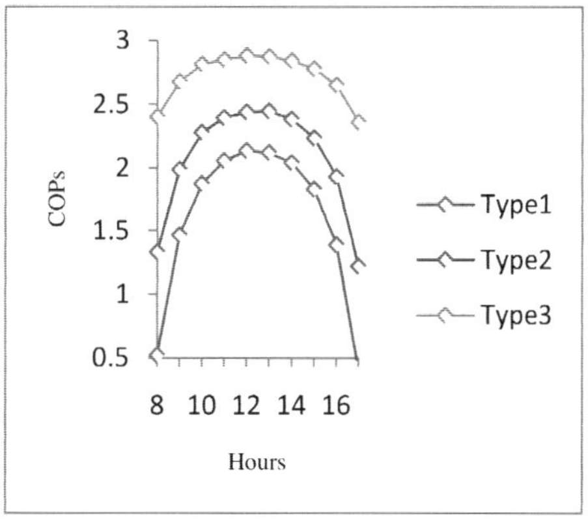

Figure 10. Hourly variation of the COP_s for three flat plate collectors

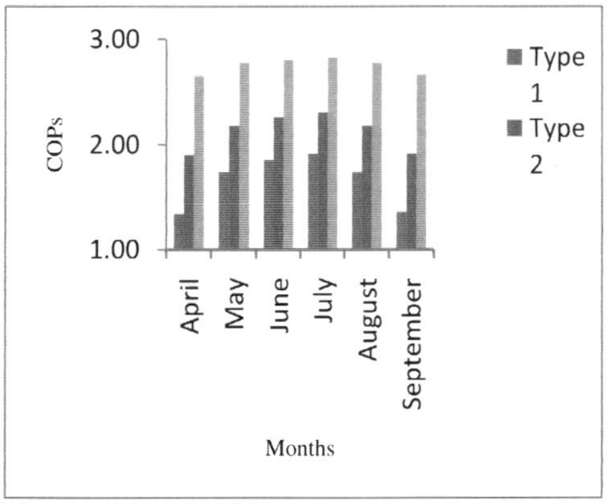

Figure 11. Monthly average performance of the solar combined ejector system for the three flat plate collector types

6. Conclusion:

The main objective of this work was to analyse a combined ejector-vapour compression system assisted by solar energy. The system analysis was based on the empirical correlation and friendly working fluid.

The authors believe that the analysed system is well posed to offer a good answer to the increased demand for refrigeration and air-conditioning. The COP of the combined ejector cycle is mainly influenced by the type and the efficiency of the solar collector. For the considered refrigerants, it has been observed for a given operating conditions that the best performances are obtained with R290. Even though, is necessary to investigate the performance of the system under different conditions from the ones considered in the present study like the use different design parameters.

References:

[1] Zeren. F, Holmes. RE, Jenkins. PE, 1979. Design of Freon jet pump for use in solar cooling system.ASME Paper No. 78 W A SOL-15

[2] Huang.BJ, Chang. JM, Petrenko. VA, Zhuk. KB, 1998. A solar ejector cooling-system using refrigerant R141b. *Sol. Energ.* **64 (4-6)** 223-226

[3] Nahdi. E, Kairouani. L, Elakhdar. M, 2008. A solar ejector air conditioning system using environment-friendly working fluids. *Int. J. of Energ. Res.* **32** 1194-1201

[4] Sokolov. M, Hershgal. D, 1990. Enhanced ejector refrigeration system cycles powered by low grade heat. Part 1: systems characterization. *Int. J. of Refrig.* **13 (6)** 351-356

[5] Hernandez, J.I., Dorantes, R.J., Best, R., Estrada, C.A., 2004. The behaviour of a hybrid compressor and ejector refrigeration system with refrigerants 134a and 141b. *Appl. Ther. Eng.* **24(5)** 1765-1783

[6] Vidal. H, Colle. S, 2010. Simulation and economic optimization of a solar assisted combined ejector-vapor compression cycle for cooling applications. *Appl. Ther. Eng.* **30** 478-486

[7] Sun. DW, 1997. Solar powered combined ejector-vapour compression cycle for air conditioning and refrigeration. *Energ. Conv. Manag.* **38(5)** 479–791

[8] Duffie. JA, Beckman. WA, 1980. Solar Engineering of Thermal Process. Wiley/Inerscience: New York

[9] Pridasawas.W, Lundqvist. P, 2007. A year-round dynamic simulation of a solar-driven ejector refrigeration system with iso-butane as a refrigerant. *Int. J. of Refrig.***30** 840-850

[10] Nahdi. E, Champoussin. JC, Hostache. G, Cheron. J, 1993. Optimal geometric parameters of a cooling ejector-compressor. *J. Int. De Froid* **16** 67-72

[11] Holton. WC, Schultz. EJ, 1951. Effect of molecular weight of entrained fluid on the performance of steam jet ejector. *Transactions of ASME* **73(7)** 905-910

[12] Work. LT, Haedrich. VW, 1939. Performance of ejectors as function of the molecular weights of vapours. *Indust. and Eng. Chem* **31** 464-477

[13] Lu. LT, 1996. Etude théorique et expérimentale de la production du froid par machine tritherme à éjecteur de fluide frigorigène. Thèse de doctorat. Institut National Polytechnique de Grenoble.

[14] Dorantes. R, Lallemand. A, 1995. Influence de la nature des fluides, purs ou en mélanges non-azéotropiques sur les performances dune machine de climatisation à éjecto-compresseur. *Int. J. of Refrig.* **18** 21-30

[15] Huang. BJ, Petrenko. VA, Chang. JM, Lin. CP, Hu. SS, 2001. A combined cycle refrigeration system using ejector cooling cycle as the bottom cycle. *Int. J. of Refrig.***24** 391-9

[16] REFPROP (2007) Thermodynamic Properties Refrigerant Mixtures, Version 8.1. National Institute of Standards and Technology, 2002

[17] Institut National de Météorologie. Données climatiques de Tunis

Possibility of Si-based new material for thin-film solar cell applications

T Suemasu
Institute of Applied Physics, University of Tsukuba, 1-1-1 Tennohdai, Tsukuba, Ibaraki 305-8573, Japan.

E-mail: suemasu@bk.tsukuba.ac.jp

Abstract. We grow $BaSi_2$ epitaxial films on Si(111) substrates by molecular beam epitaxy, and investigate their optical properties such as optical absorption coefficients, minority-carrier diffusion length, and minority-carrier lifetime. These are key parameters which determine the solar cell performance. The band gap of $BaSi_2$ is measured to be approximately 1.3 eV. The absorption coefficient reaches approximately 3×10^4 cm^{-1} at 1.5 eV. The minority-carrier diffusion length and minority-carrier lifetime are found to be about 10 μm and 8 μs, respectively. These values are sufficiently large for thin-film solar cell applications. Internal photoresponse spectra are deduced from data obtained experimentally.

1. Introduction

Recently photovoltaic solar cells have been attracting considerable attention. At present, approximately 90% of solar cells are based on crystalline Si because of its earth-abundance and developed technologies.Thin-film solar cell materials such as CIGS, CZTS, CdTe, and organic materials have been attracting increasing attention due to their high efficiency and low cost. In contrast, the optical absorption layers of crystalline Si solar cells tend to be much thicker than conventional thin-film solar cells, because the optical absorption coefficient α is much smaller for crystalline Si. Therefore, novel Si-based materials for high-efficiency thin-film solar cells have received significant interest. However, little steadfast effort has been devoted to any materials other than Si, CZTS, CIGS, CdTe and III-V compounds as far as inorganic semiconductors are concerned. Among such materials, we have focused much attention on semiconducting $BaSi_2$. The band gap of $BaSi_2$ is approximately 1.3 eV [1,2], and can be increased up to 1.4 eV in $Ba_{1-x}Sr_xSi_2$[3], which matches the ideal solar spectrum much better than crystalline Si. In addition, $BaSi_2$ has a very large absorption coefficient of approximately 3×10^4 cm^{-1} at 1.5 eV [2]. A large value of α and expansion of the band gap in $Ba_{1-x}Sr_xSi_2$ were theoretically expected [4,5]. In solar cells, both absorption coefficient α and minority-carrier diffusion length L are critical. In order to efficiently extract photogenerated carriers, the diffusion length should be larger than the absorption length. Direct band gap semiconductors like GaAs usually have large absorption coefficient. But this generally comes with faster recombination and hence relatively smaller diffusion length. In contrast, both α and L were found to be large enough in $BaSi_2$ (L~10μm) [2,6]. This is because $BaSi_2$ is an indirect band gap semiconductor with a direct transition 0.1 eV above the band edge. Recently, we successfully achieved large photoresponsivity and internal quantum efficiency exceeding 70% in 0.4μm-thick *a*-axis-oriented $BaSi_2$ epitaxial layers [7].

Excess carrier recombination mechanisms in $BaSi_2$ have also been studied [8].These results have spurred interest in this material.

Control of the conductivity of $BaSi_2$ by impurity doping is a requirement because the basic structure of a solar cell is a *p-n* junction. According to Imai and Watanabe [9], substitution of Si in the $BaSi_2$ lattice is more favorable than substitution of Ba from an energetic point of view by first-principles calculation. In our previous works, the electron concentration of Sb-doped n-type $BaSi_2$ was controlled in the range between 10^{16} and 10^{20} cm^{-3} at room temperature (RT) [10]. Very recently, we have achieved the hole concentration exceeding 10^{19} cm^{-3} in B-doped p-type $BaSi_2$. The hole concentration of B-doped $BaSi_2$ was controlled in the range between 10^{16} and 10^{20} cm^{-3} at RT. In this article, the current status and future prospect of this material towards high-efficiency and thin-film solar cell applications are reviewed.

2. Experimental procedures

An ultrahigh vacuum (UHV) chamber equipped with a Knudsen cell for Ba and an electron beam gun for Si was employed. Before the growth, the Si(111) substrates were prepared by subjecting them to the following treatment. The substrates were washed using RCA clean steps, which removed organic and metallic contaminants. The substrates were then annealed at 830 °C for 30 min in the UHV (1×10^{-6} Pa) chamber to remove the protective SiO_2 layers. After annealing, a 7×7 streaky reflection high-energy electron diffraction (RHEED) pattern was observed, indicating a clean Si surface. A two-step growth method was adopted, which included reactive deposition epitaxy (RDE; Ba deposition on hot Si) and molecular beam epitaxy (MBE; co-deposition of Ba and Si) to form thick $BaSi_2$ films [11,12]. The RDE process was carried out for deposition of a template layer as a $BaSi_2$ precursor prior to the subsequent MBE process. The same growth method was successfully utilized for the epitaxial growth of semiconducting β-$FeSi_2$ films on both Si(001) and Si(111) substrates [13,14]. The crystalline quality of the films was evaluated using RHEED and θ-2θ X-ray diffraction (XRD) measurements using Kα x-rays.

3. Results and discussions

3.1. Optical absorption properties

For measurements of optical absorption coefficients, we fabricated the silicon-on-insulator (SOI) substrates using wafer bonding at room temperature (RT) as well as chemical mechanical polishing (CMP). First, 500-μm-thick high-resistive FZ-Si(111) ($\rho > 1000$ Ω·cm) and 500-μm-thick fused silica wafers were bonded at RT. Then, the Si wafer was mechanically ground and polished by CMP down to about 0.7 μm thickness. Prior to loading the substrates into the MBE chamber, oxide layers on the surface of the SOI substrate were etched away by diluted hydrofluoric acid. Thermal treatment was performed at 590 °C for 20 min in UHV to clean the hydrogen-terminated surface. A 20-nm-thick $BaSi_2$ template layer was then formed at 525 °C using RDE, prior to the deposition of an 80-nm-thick $BaSi_2$ layer by MBE (sample A). Si and Ba atoms were co-deposited at 575°C by an electron-beam gun and a Knudsen cell, respectively. For comparison, $BaSi_2$ layers were also epitaxially grown on a Si(111) substrate (sample B) in the same manner as described previously [12]. RHEED and θ-2θ XRD patterns were utilized to evaluate the crystalline quality of the $BaSi_2$ layers. We employed a JASCO U-best 570 spectrophotometer to obtain the transmission spectra of the samples. We observed the (1×1) streaky pattern of Si(111) after thermal cleaning at 590°C.

Figures 1(a) and 1(b) show RHEED patterns of the RDE-grown $BaSi_2$ template layer. The incident electron beam was along the (a) Si[1-10] and (b) Si[11-2] directions. After the RDE growth, we confirmed that the $BaSi_2$ film was epitaxially grown on the Si layer of the SOI substrate as shown in Figs. 1(c) and 1(d). Since these patterns could also be observed at the end of MBE growth, we determined that the $BaSi_2$ film was indeed grown epitaxially.

Figure 2 shows the θ-2θ XRD patterns of samples A and B, $BaSi_2$ films grown on the SOI and Si(111) substrates, respectively. We can see diffraction peaks only from (100)-oriented $BaSi_2$ planes, such as the (200), (400), (600) planes. For the $BaSi_2$ epitaxial film on the SOI substrate, the peak intensities of these planes were almost the same as those in the $BaSi_2$ epitaxial film on the Si(111) substrate. These results indicate that the crystalline quality of the $BaSi_2$ film on the SOI substrate was equivalent to that on the single-crystalline Si(111) substrate.

The transmission spectrum for the $BaSi_2$/SOI structure is presented in Fig. 3. This spectrum was significantly influenced by an interference effect within the 0.7-μm-thick Si layers. The Si layers were very thin and flat due to the CMP process, so that these interference fringes were superimposed on the spectrum. We used the equations presented in Ref. 15 to derive the interference-free transmission spectrum. As shown in Fig. 3, the transmission spectrum was fitted by the maximal extremes of the interference fringes (T_M) and also by their minimal extremes (T_m). Assuming that the Si layers have a certain absorption coefficient on a transparent substrate, the interference-free transmission spectrum T_α can be expressed as just the geometric mean of T_M and T_m ($T_\alpha = \sqrt{T_M T_m}$) over the entire region of the transmission spectrum. We calculated the T_α values using the T_M and T_m curves, and also eliminated the absorption due to the Si layers by measuring the transmission spectrum of another SOI substrate. We finally normalized T_α by taking the reflectivity of the sample into consideration. Using T_α, we could obtain the absorption spectrum shown in Fig. 4(a). The absorption coefficient of $BaSi_2$ reached 3×10^4 cm^{-1} at 1.5 eV. Figure 4(b) shows the $(\alpha dh\nu)^{1/2}$ versus hν plot for deriving the indirect optical absorption edge. The straight fit line intersects the horizontal axis at 1.34 eV. Thus, the indirect absorption edge with phonon emission was 1.34 eV. The absorption with phonon absorption is neglected, because the fitting was performed at the higher-energy region than the indirect band gap. Thus, the indirect band gap is determined to be approximately 1.32 eV, assuming that the phonon energy is approximately 25 meV. It is true that absorption coefficients below 10^2 cm^{-1} are necessary to determine the precise energy gap. Much thicker $BaSi_2$ enables us to measure smaller absorption coefficients free from errors. In addition, we can't rule out the possibility that defects contribute to absorption to some extent in the energy range below 1.34 eV.

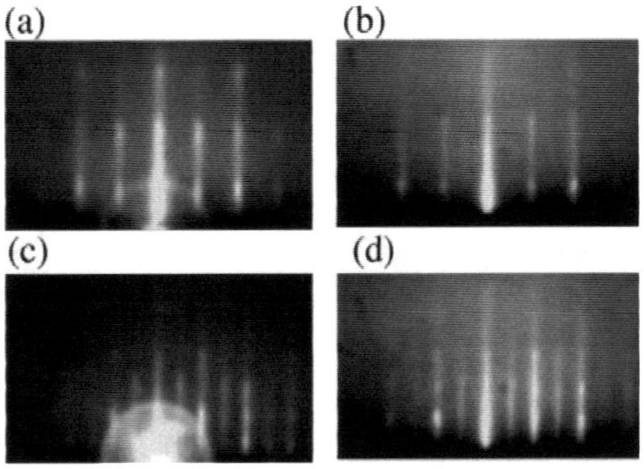

Figure 1. RHEED patterns of $BaSi_2$ in sample A, observed after (a)(b) RDE and (c)(d)MBE growth along the (a)(c) Si[1-10] and (b)(d) Si[11-2] azimuths.

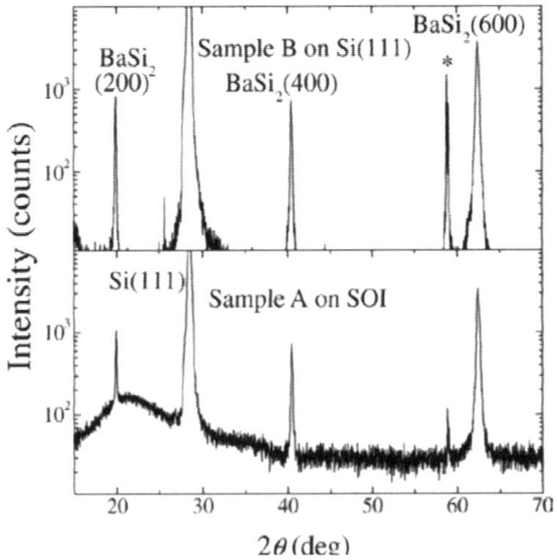

Figure 2. θ-2θ XRD patterns for samples A and B, grown on SOI and Si(111) substrates, respectively. The forbidden diffraction peak of Si(222) indicated by asterisk occurs by double diffraction.

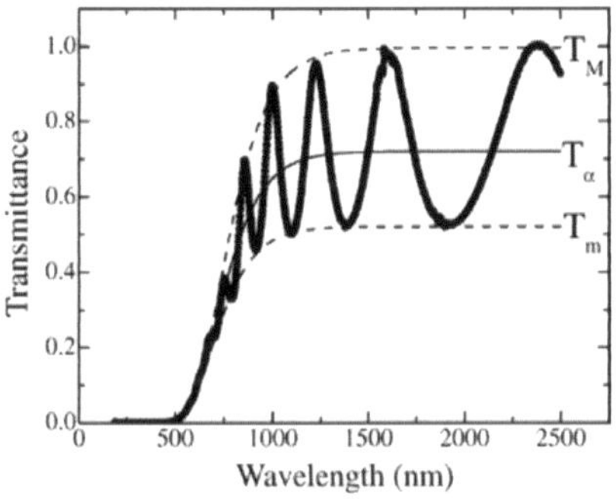

Figure 3. Transmission spectrum for sample A, where $T_\alpha = \sqrt{T_M T_m}$.

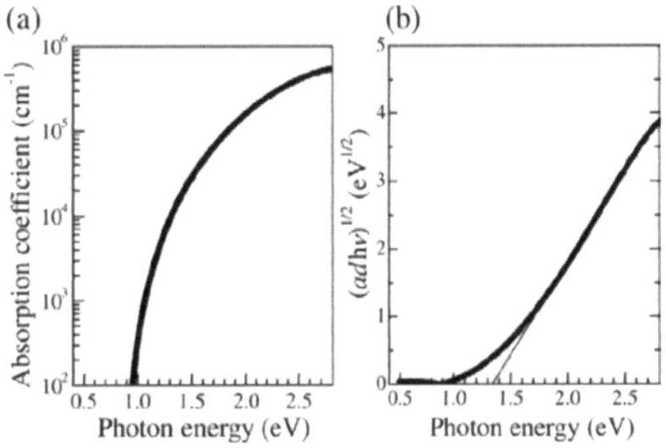

Figure 4. (a) Absorption spectrum for sample A. (b) $(\alpha dh\nu)^{1/2}$ versus $h\nu$ plot for sample A.

3.2. Minority-carrier properties

3.2.1. Minority-carrier diffusion length

For measurements of minority-carrier diffusion length in undoped $BaSi_2$, we fabricated 300-nm-thick undoped n-$BaSi_2$ films by RDE at 550 °C for 5 min, followed by MBE at 600 °C for 120 min. Undoped $BaSi_2$ shows n-type conductivity with electron concentrations of approximately 10^{16} cm^{-3} [1]. In order to investigate the grain size of $BaSi_2$ and grain boundaries (GBs), plan-view transmission electron microscopy (TEM) samples prepared by mechanical polishing and ion milling were observed using TOPCON EM-002B operated at 120 kV. For electron beam induced current (EBIC) measurements, Al/n-$BaSi_2$ Schottky diodes were formed. Front-side Schottky contacts were formed with Al on the $BaSi_2$ surface via wire bonding, and the back-side ohmic contact was made with Al by sputtering. EBIC observations were carried out in the edge-scan configuration with a Hitachi S4300 field-emission scanning electron microscope (SEM) in the EBIC mode at RT. The acceleration voltage of the electron beam, V_{ac}, was set at 5 kV to avoid penetration of the beam into the Si substrate. The penetration depth of the electron beam is estimated to be shorter than 300 nm, which is the thickness of the $BaSi_2$ layers, when V_{ac} is 5 kV, with the density of $BaSi_2$ being 5.14 g/cm^3.

Figures 5(a)(b) and 5(c)(d) show secondary-electron (SE) and EBIC images around the Al contact, respectively, with V_{ac} = 5 kV. In the EBIC method, carriers generated within the diffusion length in the n-type $BaSi_2$ are collected by the electric field under the Al contact and sensed as a current in the external circuit. In Figs. 5(c)(d), the brighter regions show higher collection of electron-beam-induced carriers in the $BaSi_2$. We cannot see defect-related black lines. Figure 6 shows the EBIC line-scan data along dotted line AA' in Fig. 5(c). The EBIC profile shows an exponential dependence of the distance from the Al contact. In this work, the diffusion length of minority carriers was roughly estimated to be approximately 10 μm, assuming that the EBIC profile varies as $\exp(-x/L)$, where x is the distance from the Al edge (point A) along the dotted line, and L is the diffusion length of holes for $BaSi_2$. The obtained minority-carrier diffusion length is much larger than the grain size of the $BaSi_2$, implying that the GBs do not work as defect centers for minority carriers in n-$BaSi_2$. The contribution of carriers generated within the n-Si substrate to the measured EBIC signals can be excluded, because the simulated penetration depth of the electron beam is shorter than 300 nm, the thickness of the $BaSi_2$ layers, when V_{ac} is 5 kV. Thus, it is reasonable to think that the number of carriers generated within the Si substrate was negligibly small compared to those generated in the $BaSi_2$ layers. To confirm the GBs character, an EBIC system with spatial resolutions less than 0.1 μm may be necessary. We should also note here that the obtained minority-carrier diffusion length is roughly 30 times larger than $1/\alpha$ (=0.3 μm) at 1.5 eV, suggesting that this value is large enough for solar cell applications.

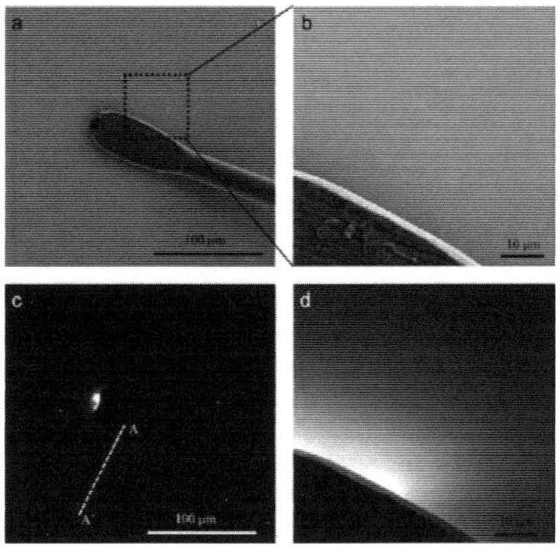

Figure 5. (a)(b) SE and (b)(d) EBIC images around the Al contact.

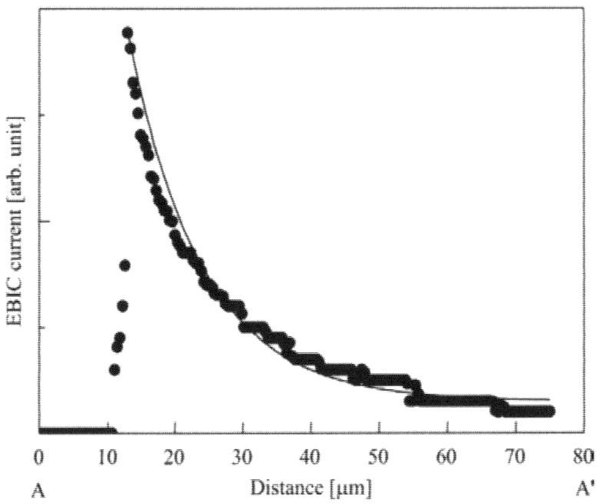

Figure 6. Experimental and simulated (solid line) EBIC line-scan profiles along the dotted line from points A to A' in Fig. 5(c).

3.2.2. Minority-carrier lifetime diffusion length

After cleaning a floating-zone n-Si(111) substrate (ρ>1000 Ω·cm) at 900 °C for 30 min in UHV, a 5-nm-thick $BaSi_2$ template layer was grown on the Si substrate by RDE, followed by MBE to form a 1.5-μm-thick $BaSi_2$ epitaxial film. Figure 7 shows the relationship between apparent minority-carrier life time τ_a and excess-carrier density of the 1.5-μm-thick $BaSi_2$. The undoped $BaSi_2$ shows n-type conductivity and the electron concentration is about 10^{16} cm^{-3} [1]. The excess carrier concentration was calculated from the absorption coefficient of $BaSi_2$ at 349 nm, and the irradiated laser intensity, which was varied from 1.1×10^2 to 1.3×10^2 W/cm^2. The inset shows the decay curve when the excess carrier concentration was 7.4×10^{16} cm^{-3}. Decay can be divided into two modes in terms of decay rate. We see initial rapid decay, followed by approximately constant decay. According to our previous work, the rapid decay is caused by Auger recombination [8]. τ_a was calculated by approximating the tail region of the decay curve by a single exponential curve. We see that τ_a decreases with increasing the excess-carrier density. This is because the excess-carrier density is several orders of magnitudes higher than majority-carrier density at equilibrium in undoped n-$BaSi_2$. Such high carrier injection leads to multicarrier recombination. Therefore, when we discuss about the minority-carrier lifetime, we should use lower injection value. As can be seen in Fig. 7, τ_a becomes independent of excess-carrier density when the excess carrier concentration becomes smaller than 10^{16} cm^{-3}, indicating that τ_a corresponds to the effective minority-carrier lifetime τ in this region. This result makes sense because the electron concentration of undoped n-$BaSi_2$ is about 10^{16} cm^{-3} [1]. Therefore, it

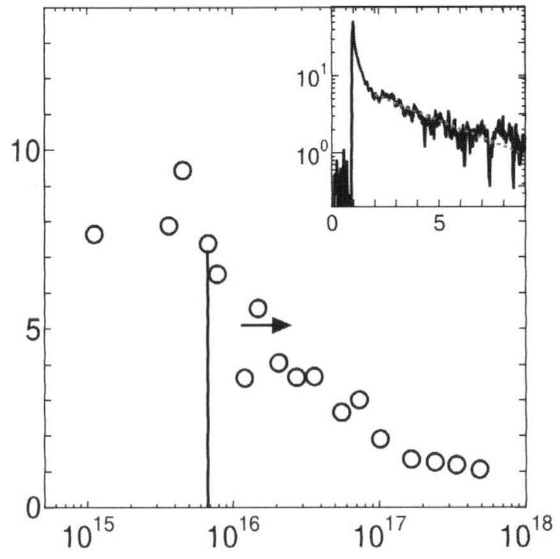

Figure 7. Relationship between minority-carrier life time and excess-carrier density of 1.5-μm-thick $BaSi_2$. The inset shows the photoconductivity decay curve of the 1.5-μm-thick $BaSi_2$ film when the excess-carrier concentration is 7.4×10^{16} cm^{-3}.

is reasonable to think that the minority-carrier lifetime in the 1.5-μm-thick $BaSi_2$ is about 8 μs. Assuming that the hole mobility is 1 $cm^2V^{-1}s^{-1}$ [16], the minority-carrier diffusion length L is estimated to be about 4.5 μm from the equation of $L=\sqrt{k_B T\mu\tau/q}$. Here, k_B is the Boltzmann constant, T the absolute temperature, μ the mobility of holes, τ the effective minority-carrier lifetime, and q the elemental charge. This value is about three times larger than the absorption layer thickness (~1.5 μm) in a $BaSi_2$ pn junction diode. On the basis of these results, we achieved the formation of high-quality $BaSi_2$ films thicker than 1.5 μm, where the minority-carrier lifetime was large enough for thin-film solar cell applications.

3.3. Expected spectral response in a $BaSi_2$ pn junction diode

We attempted to deduce the expected photocurrent density in a $BaSi_2$ p^+n homojunction diode. Analytical expressions for the photogenerated and dark saturation current densities can be obtained only when the dopant concentration, carrier lifetime and other parameters are assumed constant. Suppose here a simple p^+n abrupt junction diode in samples pn-A to pn-E as shown in Table 1. N_A, W_e, L_e, S_e are the acceptor density, neutral p^+ layer width, minority-carrier (electrons) diffusion length, and surface recombination velocity, and N_D, W_b, L_h, and S_h are their counterparts in the n layer. W is the depletion region width.

Table 1. Parameters of $BaSi_2$ homojunction diode are shown.

Sample	N_A (cm^{-3})	W_e (μm)	L_e (μm)	S_e (cm/s)	W (μm)	N_D (cm^{-3})	W_b (μm)	L_h (μm)	S_h (cm/s)
pn-A	10^{18}	0.1	0.1	8	0.4	10^{16}	1.5	10	8
pn-B	10^{18}	0.1	0.1	8	0.4	10^{16}	2.5	10	8
pn-C	10^{18}	0.05	0.1	8	0.4	10^{16}	1.5	10	8
pn-D	10^{18}	0.05	0.1	100	0.4	10^{16}	1.5	10	8
pn-E	10^{18}	0.05	0.1	8	0.4	10^{16}	1.5	10	100

The total photocurrent density j at energy E per unit energy is given by the sum of current due to electrons generated within the neutral p^+-$BaSi_2$ region, j_e, current due to electron-hole pairs generated within the depletion region, j_{DL}, and current due to holes generated within the neutral n-$BaSi_2$ region, j_h. The internal quantum efficiency, IQE, is equal to,

$$IQE(E)=\frac{j}{q\Phi_0(1-R)}=\frac{j_e+j_{DL}+j_h}{q\Phi_0(1-R)} \qquad (1)$$

Once the IQE is derived, the J_L is given by,

$$J_L=q\int\Phi_0(1-R)\,IQE(E)\,dE \qquad (2)$$

In order to make J_L large, we should minimize R and maximize IQE over the solar spectral range. Therefore, the IQE is a measure of the quality of the diode. The contribution to the IQE by j_{DL}, j_e and j_h, that is IQE_{DL}, IQE_e, and IQE_h, respectively, are obtained analytically. Then, we investigated the dependence of IQE on various parameters. Regarding the α, we used the experimental data shown in ref 1. While the diffusivity is not highly sensitive to impurity and defect, we thus set both D_e and D_h to be 0.1 cm^2/s, because L_h is 10 μm and τ_h is 10 μs as described earlier. When the impurity concentrations in the p^+ and n layers, that is N_A and N_V, are set to be 10^{18} and 10^{16} cm^{-3}, respectively. In this case, the W is estimated to be 0.4 μm and the built-in voltage 1.0 V assuming that the permittivity of $BaSi_2$ is approximately 15, and the effective densities of states of the conduction band

and valence band, that is, N_C and N_V, respectively, are 2.6×10^{19} and 2.0×10^{19} cm^{-3}.[4] The square of the intrinsic carrier concentration, n_i, at RT is calculated to be $n_i^2 = 8\times10^{16}$ cm^{-3}. Other parameters are summarized in Table 1. In Figs. 8(a) and 1(b), the *IQE* is almost the same between samples pn-A and pn-B although the W_b is increased from 1.5 μm in sample pn-A to 2.5 μm in sample pn-B, meaning that W_b=1.5 μm is enough. We see that when the W_e is decreased from 0.1 to 0.05 μm, the *IQE* is increased particularly in higher energies in samples pn-C to pn-E. This is attributed to the enhancement of IQE_e and IQE_{DL}. Since the absorption coefficient of $BaSi_2$ is large, the contributions of the surface p^+-layer and the depletion region to the *IQE* becomes pronounced when the W_e is decreased. This result implies that we design the W_b to be equal to L_e or smaller. In samples pn-C to pn-E, there is no significant difference in *IQE* regardless of different S_e and S_h. This means that the *IQE* is not so sensitive to surface conditions in a $BaSi_2$ homojunction solar cell. We next discuss the expected conversion efficiency of the diodes. Considering that n_{p0} is much smaller than p_{n0} because of the p^+n junction, the dark current density of the diode, J_{dark}, is dominant from the contribution of n-type layer, and thus, the reverse saturation current density J_s is obtain. The J_s and J_L values, and the expected open-circuit voltage, V_{OC}, under AM1.5 illumination for samples pn-A to pn-E are investigated, assuming that the fill-factor, *FF*, is 0.8 regardless of V_{OC} and *R*=0. As a result, the conversion efficiency is calculated to be more than 25% in samples pn-C and pn-D even though the total layer thickness is only 2 μm.

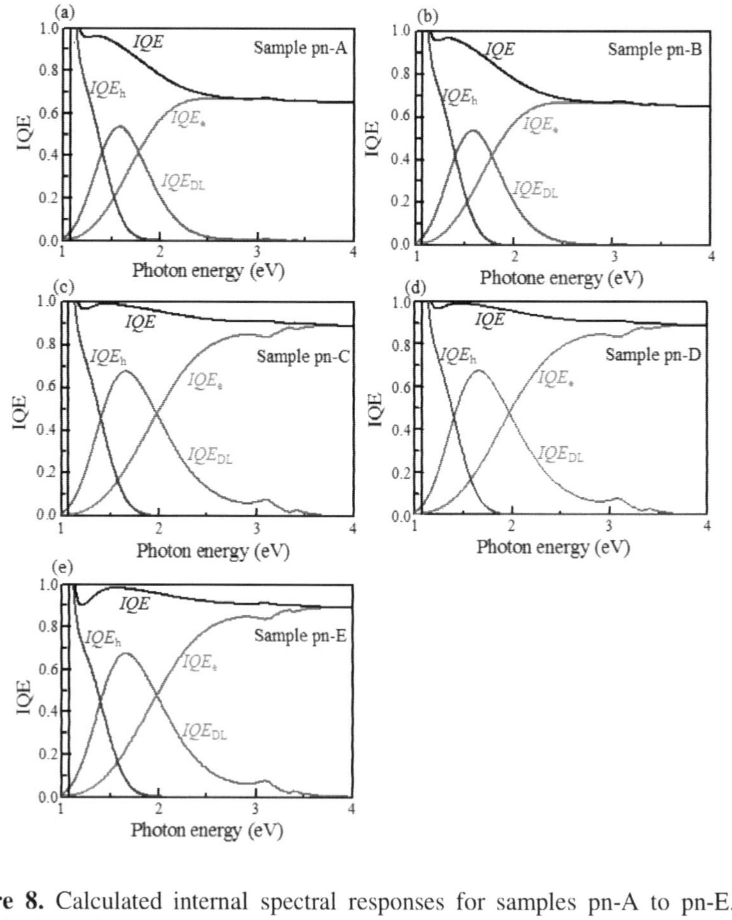

Figure 8. Calculated internal spectral responses for samples pn-A to pn-E. Individual contributions of IQE_e, IQE_{DL}, and IQE_h to *IQE* are shown.

4. Summary

Current status toward Si-based high-efficiency thin-film solar cells was reviewed in terms of optical absorption properties and minority-carrier properties. These are key parameters which determine the performance of solar cells. The band gap of $BaSi_2$ was measured to be approximately 1.3 eV. The absorption coefficient reached approximately 3×10^4 cm^{-1} at 1.5 eV. The minority-carrier diffusion length and minority-carrier lifetime were found to be about 10 μm and 8 μs, respectively. In particular, what's interesting with $BaSi_2$ is that it has large absorption coefficients and at the same time large minority-carrier diffusion length. This feature, caused by the fact that the direct transition starts just 0.1 eV above the band edge, is favourable to effectively extract photogenerated carriers. We calculated the photoresponse spectra of $BaSi_2$ p^+n homojunction diodes using experimentally obtained absorption coefficients and minority-carrier diffusion length, and discussed the influence of parameters such as emitter and base layer thicknesses, surface recombination velocities. The *IQE* is not so sensitive to S_e and S_h, and W_b, while the W_e has a significant influence on *IQE* particularly in a high energy region. This is because the absorption coefficient of $BaSi_2$ is large. Thus, we need to take care of the relationship between W_e and L_e in the emitter layer when we design the W_e. The conversion efficiency exceeding 25% is expected for 2 μm-thick $BaSi_2$ homojunction diodes in an ideal condition.

[1] Morita K, Inomata Y and Suemasu T 2006 *Thin Solid Films* **508** 363
[2] Toh K, Saito T and Suemasu T 2011 *Jpn. J. Appl. Phys.* **50** 068001
[3] Morita K, Kobayashi M and Suemasu T 2006 *Jpn. J. Appl. Phy.* **45** L390
[4] Migas D B, Shaposhnikov V L and Borisenko V E 2007 *Phys. Status Solidi* B **244** 2611
[5] Imai Y and Watanabe A 2007 *Thin Solid Films* **515** 8219
[6] Baba M, Toh K, Toko K, Saito N, Yoshizawa N, Jiptner K, Sekiguchi T, Hara K O, Usami N and Suemasu T 2012 *J. Cryst. Growth* **348** 75
[7] Du W, Suzuno M, Khan M. A, Toh K, Baba M, Nakamura K, Toko K, Usami N and Suemasu T 2012 *Appl. Phys. Lett.* **100** 152114
[8] Hara K O, Usami N, Toh K, Baba M, Toko K and Suemasu T 2012 *J. Appl. Phys.* **112** 083108
[9] Imai Y and Watanabe A 2007 *Intermetallics* **15** 1291
[10] Kobayashi M, Matsumoto Y, Ichikawa Y, Tsukada D, and Suemasu T 2008 Appl. Phys. Express **1** 05403
[11] Inomata Y, Nakamura T, Suemasu T and Hasegawa F 2004 *Jpn. J. Appl. Phys.* **43** 4155
[12] Inomata Y, Nakamura T, Suemasu T and Hasegawa F 2004 *Jpn. J. Appl. Phys.* **43** L478
[13] Tanaka M, Kumagai Y, Suemasu T and Hasegawa F 1997 *Jpn. J. Appl. Phys.* **36** 3620
[14] Hiroi N, Suemasu T, Takakura K, Seki N and Hasegawa F 2001 *Jpn. J. Appl. Phys.* **40** L1008
[15] Swanepoel R 1983 *J. Phys.* E **16** 1214
[16] Khan M A, Hara K O, Du W, Baba M, Nakamura K, Suzuno M, Toko K, Usami N and Suemasu T 2013 *Appl. Phys. Lett.* **102** 112107

A NUMERICAL study of solar chimney power plants in Tunisia

Attig Bahar F[1], Guellouz M S[2], Sahraoui M[1] and Kaddeche S[3]

[1] University of Carthage, Polytechnic School of Tunisia, BP. 743, La Marsa. Tunisia.
[2]University of Carthage, National Engineering School of Bizerte, BP66, Campus Menzel Abderrahman, 7035 Bizerte, Tunisia.
[3]University of Carthage, National Institute of Applied Sciences and Technology, Laboratory of Materials, Measurements and Applications. BP. 676, 1080 Tunis, Tunisia.

E-mail: faten.bahar@gmail.com

Abstract. A 3D CFD (Computational fluid dynamics) model of a Solar Chimney Power Plant (SCPP) was developed and validated through comparison with the experimental data of the Manzanares plant. Then, it was employed to study the SCPP performance for locations throughout Tunisia.

Key words: solar power, solar chimney, Numerical study.

1. Introduction

The solar chimney power facility has the potential to become a valuable technology for renewable electrical energy production. However, its financial viability depends on a thorough understanding of the processes affecting its performance, particularly because of the large startup costs associated with facility design and construction.The first solar chimney prototype was built in Manzanares Spain with a chimney that has a height of 195 m and a dimater of 10 m, surrounded by a collector 240 m in diameter[1,2]. The power output produced by this plant was about 50 kW. Other prototypes were built in Algeria [3] and Iran [4]. Larger projects are planned in Australia and Spain.

High levels of direct solar radiations and large available lands in Tunisia's desert are major factors to encourage the development of solar power plants for both thermal and electrical energy production for various uses. Among these, Solar Chimney Power Plants (SCPP) are quite attractive, for their technological simplicity but mainly because they do not require cooling nor the use of water, a scarce resource in the dessert.The present work is the first step of a research program aimed at studying the performance of the solar chimney plants in the Tunisian context.

2. Methodology

The numerical simulations were performed for steady flow using the commercial CFD paçkage ANSYS Fluent [5]. They involved the simultaneous solutions of the continuity, momentum and energy equations. Given that the turbulent flow inside the chimney power plant system is induced because of the density variation with tempertaure, the RNG k-ε turbulence model including buoyancy effects (Boussinesq approximation) was considered. The Simple algorthim was selected as the pressure-velocity coupling scheme and the convective terms were discretized with a first order upwind scheme. The Presto!algorithm was selected for the treatment of the pressure in the momentum equation. Structured grid was adopted and it was adaptively refined in the vicinity of the walls.

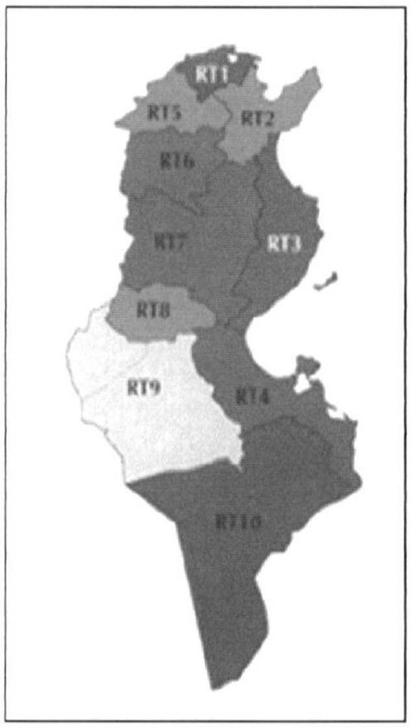

Figure 1. Locations of 10 selected regions for solar chimney power plant (SCPP) performance evaluation in Tunisia [6].

The SCPP is presented schematically in Figure 2. The main aerodynamic boundary conditions are listed in Table 1. The thermal boundary conditions are considered adiabatic except for the surface of the collector, where an incoming radiation is imposed, and the ground, where the constant soil temperature at a depth of 5 m is used as boundary condition. This temperature is obtained by solving the space-time heat equation in the soil during a year.

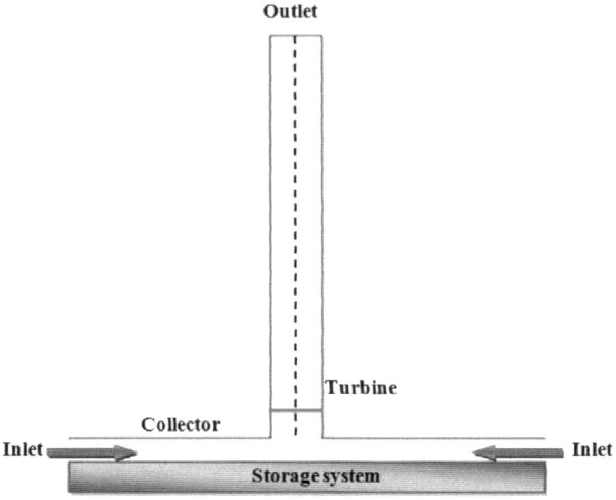

Figure 2. The Solar chimney plant schematic layout.

The solar Load model provided by Fluent was adopted to estimate this solar radiative heat. The latter was modeled using the sun position vector and the illumination parameters. Such a model is an efficient and practical approach to applying solar loads as heat sources in the energy equations.

Table.1. The thermal and aerodynamic boundary conditions for the SCPP simulations.

Boundary conditions	**Type**
Surface of the chimney	Wall, adiabatic
Surface of the collector	Mixed semi-transparent wall, (radiation, convection)
Inlet of the collector	Pressure inlet, (P gage=0Pa, Tinlet)
Outlet of the chimney	Pressure outlet (Pgage= 0Pa)
Bottom of the ground	Wall (T= Tground)
Pressure drop across the turbine	Reversed Fan

The numerical method was validated using the experimental data of the Manzanares prototype [1], [2]. In fact, as shown in Figure 3, a good agreement is obtained between the numerical results and the experimental data.

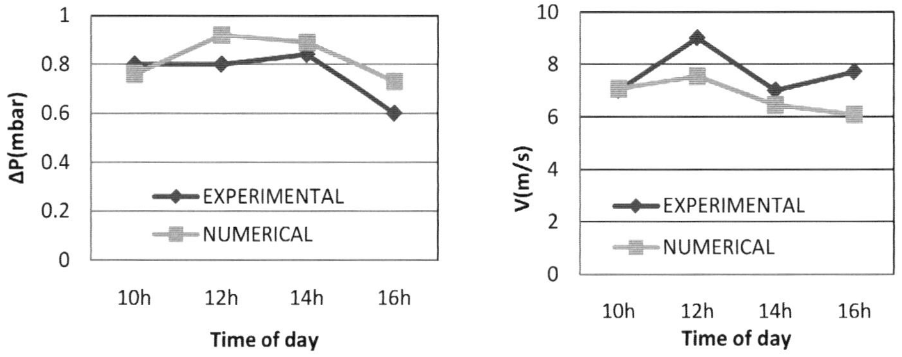

(a) (b)

Figure 3. Comparison of the calculated and experimental (a) pressure drop across the turbine and (b) outlet velocities, for September 2, 1989.

3. Numerical and Theoretical Results

As a first step in the study of the SCPP performance in the Tunisian context, we've considered geometrically similar plants to that of Manzanares implemented in 10 different regions of the country. Information about these regions are presented in the Table 2. Annual solar radiation of the major selected regions is more than 5700 MJ/m². The average annual ambient temperature lies between 19°C and 23.5°C.

Table.2. Annual global solar radiation and atmospheric conditions of 10 selected regions in Tunisia [6].

Region		Longitude	Latitude	Average annual ambiant temperature (°C)	Average annual wind speed (m/s)	Annual solar radiation [MJ/m²]
Rt1	Bizerte	9°48'E	37°15'N	19.9	3.7	6168
Rt2	Tunis	10°14'E	36°50'N	20.6	4.3	6168
Rt3	Sfax	10°41'E	34°43'N	21.2	3.7	6782
Rt4	Djerba	10°46'E	33°52'N	21.8	3.3	6651
Rt5	Jendouba	8°48'E	36°29'N	20.5	2.5	5701
Rt6	Siliana	9°22'E	36°04'N	19.6	3.1	5701
Rt7	Kairouan	10°06'E	35°40'N	21.9	2.2	6691
Rt8	Gafsa	8°49'E	34°25'N	21.7	3.9	6732
Rt9	Tozeur	8°06'E	33°55'N	23.4	4.9	6733
Rt10	Rameda	10°24'E	32°19'N	23.1	2.9	6732

Figure 4 and figure 5 shows, respectively, the monthly and the annual power output from the reference SCPP at various locations, calculated with the average conditions presented in Table 2. For these preliminary calculations, the incoming solar radiation was simply estimated from the total annual value divided by the total sun hours. As expected the power output profile is proportional to the solar radiation profiles among the considered regions. Summing the results shows that the SCPP can produce from 109 to 140 MWh/year in the selected regions. It can be observed that Tozeur and Remeda have the highest power generation per year of all the studied regions.

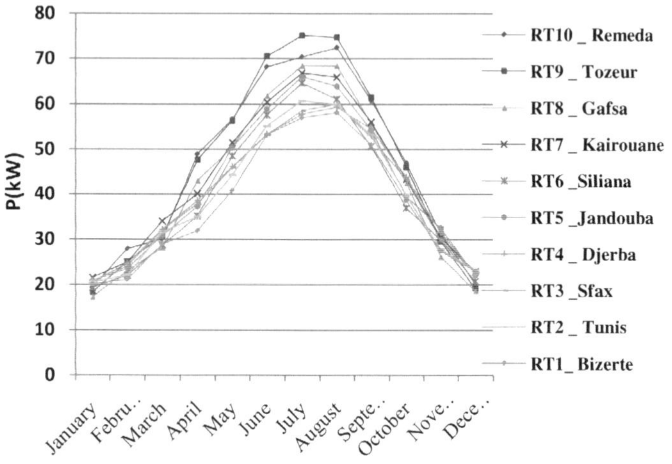

Figure 4. Monthly electrical power generation of the SCPP in selected regions of Tunisia.

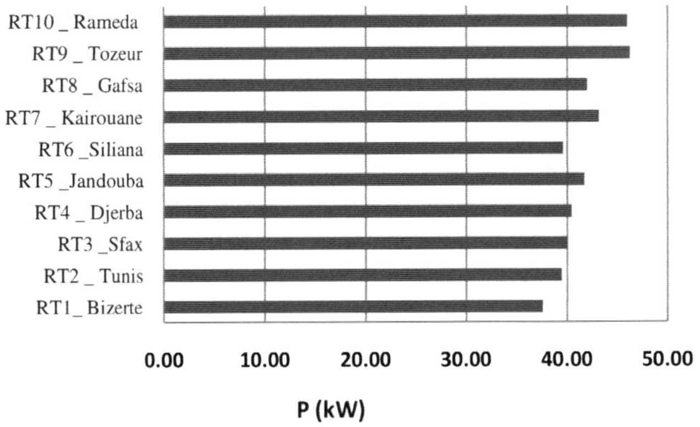

Figure 5. Annual electrical power generation of the SCPP in selected regions of Tunisia.

Figure 6 and Figure 7 show respectively the mass flow rate in the chimney and the power output of the system at the locations of Remada and Tozeur. The maximum mass flow rate and power output are observed in june, july and august when sunshine intensity and ambient temperature are at their maxima.

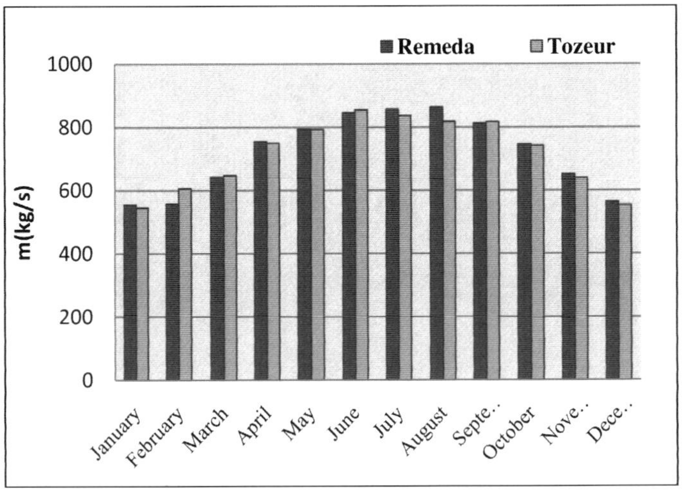

Figure 6. Mass flow rate at the chimney exit at two locations in Tunisia.

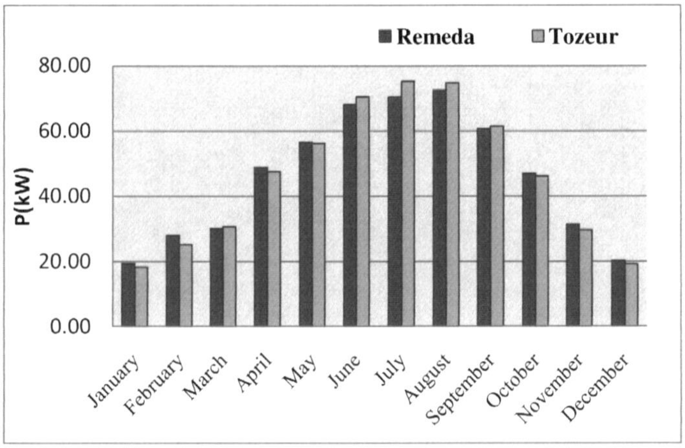

Figure 7. Monthly average SCPP power generation at two locations in Tunisia.

These results show that the solar chimney power plant can produce monthly from 4 MWh (for the month of january) to 25 MWh (for the month of july) of electrical energy. Examples of daily variations of the updraft velocity, the pressure drop across the turbine, the SCPP electrical power generation and the collector efficiency in Tozeur and Remeda were calculated for the 15th July and the 1st January. The atmospheric conditions, needed for these calculations, are given in Table3.

Table 3. Atmospheric conditions in Tozeur and Rameda for July 15th and the 1st of January.

	1st January		15thJuly		1st January		15thJuly	
	Tozeur				Remeda			
	Ta	V(m/s)	Ta	V(m/s)	Ta (°C)	V(m/s)	Ta(°C)	V(m/s)
10h00	8.68	1.86	31	8.35	11,02	2,17	27,89	3,17
12h00	11.19	1.3	32.85	8.94	12,79	2,24	32,33	0,89
14h00	12.8	3.57	36.17	2.24	15,02	2,24	33,99	4,92
16h00	13.89	2.74	37.12	4.27	14,96	1,77	33,98	4,82

The updraft velocity and the pressure drop across the turbine are presented in Figure 8 and Figure 9, for plants located in Remeda and Tozeur. Typical days are shown, corresponding to the hot (July 15) and cold (January 1) seasons. The power output of the turbine, shown in Figure 10 is directly proportional to the pressure drop and the volume flow rate across it. For the hot season, higher values of updraft velocities and pressure drop are obtained for both locations.

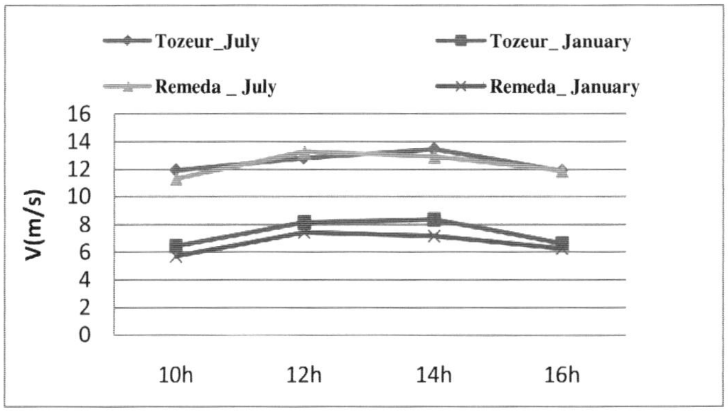

Figure 8 . The SCPP updraft velocity in m/s, for plants located in Tozeur and Remeda for the hot (July15[th]) and cold (1st of January) seasons.

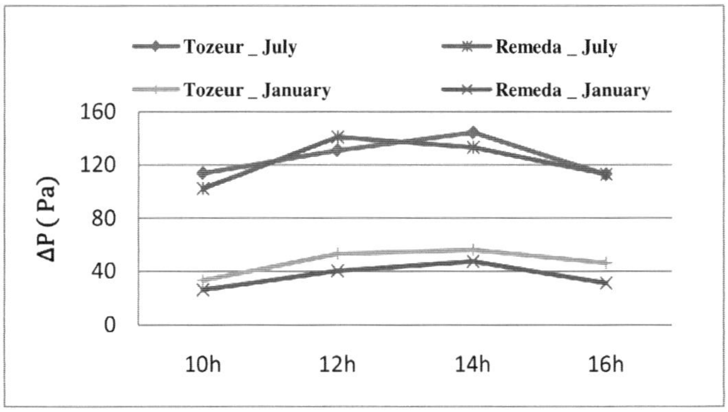

Figure 9 . The SCPP pressure drop, in Pa, for plants located in Tozeur and Remeda for the hot (July15[th]) and cold (1st of January) seasons.

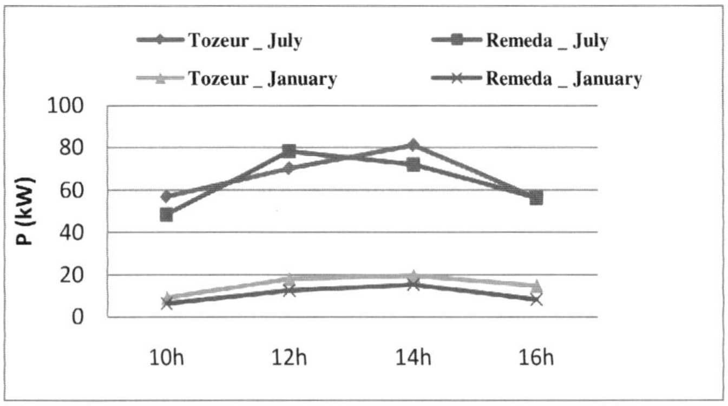

Figure 10. The SCPP electrical power generation, in kW, for plants located in Tozeur and Remeda for the hot (July15[th]) and cold (1st of January) seasons.

The collector efficiencies, defined as the ratio of heat carried by the air to the incoming solar radiation, for plants located in Tozeur and Remeda for the hot (15thJuly) and cold (1st January) seasons are shown in Figure 11 .As expected, the collector efficiency is higher during hot season, and is affected by the wind speed. In fact, it decreases with increasing wind speed as the latter increases convective losses from the collector cover.

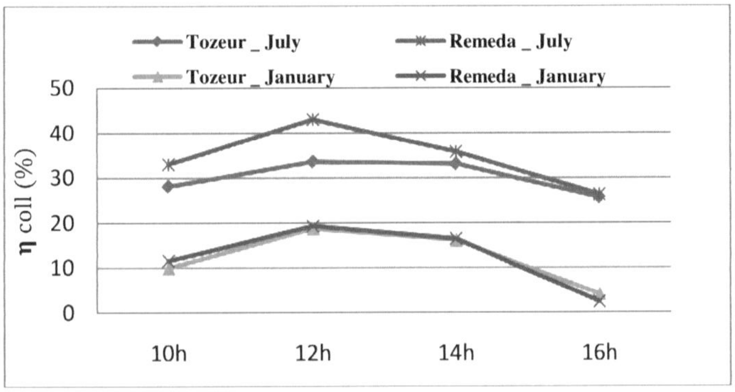

Figure 11. Collector efficiency in (%) for plants located in Tozeur and Remada for the hot (15thJuly) and cold (1st January) seasons.

4. Conclusion

A 3D CFD model of a solar chimney power plant was developed and validated through comparison with the experimental data of the Manzanares SCPP. Then, it was employed to study the SCPP performance and power generation for locations throughout Tunisia. The power generation of a solar chimney in southern regions is higher than that of other regions in Tunisia due to the higher annual solar radiation and higher sunshine duration. Particularly, Remada and Tozeur present the highest power outputs with a slight advantage to the latter. There is a vast desert land and abundant solar radiation in southern Tunisia. Therefore, these regions are suitable for the construction of SCPP.

5. References

[1] Haaf W, Friedrich K, Mayr G and Schlaich J 1983 *Int J. Solar Energy* **2** 3.
[2] Haaf W *1984 Int J. Solar Energy* **2** 141.
[3] Larbi S, Bouhdjar A and Cherguin T2010 *Renewable and Sustainable Energy Reviews* **14** 470.
[4] Asnaghi A and Ladjevardi. S.M 2012 *Energy Reviews* **16** 3383.
[5] Ansys, Fluent-solver theory guide (2011), *Ansys Inc,* USA.
[6] Agence Nationale pour la Maitrise de l'Energie, Basic climate data for the design of heating and air conditioning systems (2008), *ANME publication*, Tunisia.

Journal of Physics: Conference Series **596** (2015) 012007 doi:10.1088/1742-6596/596/1/012007

Synthesis and shell structure design of hollow silica nanoparticles using polyelectrolyte as template

M Fuji, C Takai, H Imabeppu and X Xu

Advanced Ceramics Research Center, Nagoya Institute of Technology
3-101-1 Honmachi, Tajimi, Gifu 507-0033, Japan

E-mail: fuji@nitech.ac.jp

Abstract. Hollow silica (SiO_2) nanoparticles with tunable shell microstructures were synthesized in ethanol solvent by using a polyelectrolyte as template. The polyelectrolyte template was composed of poly (acrylic acid) (PAA) with small amount of ammonia solution (NH_4OH) and prepared by dispersion of the PAA-NH_4OH mixture in ethanol. Upon the addition of silica precursor, tetraethoxysilane (TEOS), into the PAA-NH_4OH/ethanol suspension, a certain thickness of SiO_2 shell could be gradually formed outside the template through the sol-gel reaction of TEOS, which was supposed to be driven by the catalytic effect of NH_4^+ ions suspended on the surface of the template. By varying the NH_4OH amount in reaction solution, it was found that the hollow SiO_2 nanoparticles could be obtained only in the case of NH_4OH amount below a critical value. Furthermore, it was found that the water amount in reaction solution had a significant influence on the shell microstructure of hollow SiO_2. The shells formed in water-free solution were thick, rough and rich of mesopores, which became relatively thinner, smoother and richer of micropores if the water amount was gradually increased within a limited range. Therefore, it was thought to be possible for the controllable synthesis and shell structure design of hollow SiO_2 nanoparticles by optimizing the NH_4OH and water amount in reaction solution.

1. Introduction

Hollow silica (SiO_2) nanoparticles, with a unique structure of hollow interior and solid shell, can provide solid/liquid loading, thermal/electrical insulation, special optical properties, and so on [1,2]. Therefore, they have received increasing attention and have been applied in diverse fields including catalysis, drug delivery, security protection, biological detection, etc.

The current approaches to the synthesis of hollow SiO_2 nanoparticles are generally divided into two categories: (1) sacrificial template method and (2) template-free method. Of the two methods, the first one is the most effective and commonly used strategy, because it is relatively easy to control the shape and size of the resulting hollow SiO_2 nanoparticles by a simple copy surrounding the template and a subsequent removement of the template [3]. It is thus clear that the template type would be an important determining factor for the synthesis of different shapes/sizes of hollow SiO_2 nanoparticles. According to previous reports, organic templates (e.g. surfactant micelles, co-polymer micelles, etc.) mostly contributed to the formation of spherical SiO_2 hollow structure [4], while inorganic templates (compound or metal particles) might result in a shape diversity of SiO_2 hollow structure, such as rod, cubic, needle, cocoon and sphere [5,6].

The forming mechanism of the SiO_2 shell on those templates could be illustrated by either of the two following processes. One is a consecutive adsorption of pre-deposited particles on the template surface, during which the shell thickness of hollow particle was determined by the size and adsorbed amount of the pre-deposited particles [7]. The other is a precursor react, i.e. sol-gel reaction, on the template surface, where the shell thickness depended on the ratio of precursor/template [8]. In addition, the microstructure (mesopores and micropores) of the SiO_2 shell was found to be tunable by properly adjusting the sol-gel conditions, e.g. reaction time, pH value, etc [9-12]. Considering the importance of the shell thickness and shell microstructure in determining the physical and chemical properties of the hollow SiO_2 nanoparticles, it is necessary to strength the research on the two factors.

Recently, Yu and his co-workers reported a unique technique to prepare hollow SiO_2 nanoparticles using a polyelectrolyte as template [13]. The template was a mixture of poly (acrylic acid) (PAA) and ammonia solution (NH_4OH), and was insoluble in ethanol. Upon the addition of silica source (tetraethoxysilane, TEOS) into ethanol where PAA-NH_4OH template dispersed, TEOS transferred into SiO_2 shell on the template surface via a sol-gel reaction for over 10-hour stirring. A significant advantage of this method is that the template removal process is simple and environmental friendly. Although the Yu's group has showed a controllable synthesis of the hollow SiO_2 particles in particle size and shell thickness, by changing the amount of PAA and TEOS. The formation mechanism of the SiO_2 shell on the PAA-NH_4OH template was still not clear and the shell microstructure of the hollow SiO_2 particles was never mentioned. So, the purpose of our work here was mainly focused on studying the two points. First, the formation mechanism was investigated by varying the NH_4OH amount in reaction solution. Then, the shell microstructure was examined by changing the water amount in reaction.

2. Experimental section

2.1. Synthesis of hollow silica particle

Synthesis of hollow silica nanoparticles without addition of water was conducted according to Yu's groups [13]. A 0.08 g of poly (acrylic acid) (PAA, molecular weight 5,000) powder was dissolved in 1.5 ml of ammonium hydroxide solution (NH_4OH, 25 wt%) under stirring condition. The PAA-NH_4OH solution was dropped in 30 ml of ethanol (EtOH) and stirred to form PAA-NH_4OH template suspension. Then, 0.15 ml of tetraethoxysilane (TEOS) was separately added in the suspension per 2 h during 10 h stirring. After another 4 h stirring, the suspension was separated and washed by ethanol and distilled water using centrifugation. The precipitates were dried and hollow silica particles were obtained. To study the effect of NH_4OH and water content on the formation and shell microstructure of the hollow SiO_2 particles, the PAA-NH_4OH mixture was dropped in EtOH (30 ml) with addition of different NH_4OH amount (1.5, 3.0, 6.0 ml) and water amount (0.1, 1.0, 2.0 ml), respectively, and the following procedure was the same as described above.

2.2. Characterizations

Surface morphology and hollow structure of the obtained samples were observed by a scanning electron microscope (SEM), a scanning transmission electron microscope (STEM, JSM-7000F, JEOL Ltd.), and a transmission electron microscope (TEM, 2000-EX, JEOL Ltd.). Particle size of template suspension was measured by a particle size analyzer using dynamic light scattering (Zetasizer Nano, Marvern Instruments Ltd.). Specific surface areas of hollow particles were calculated by applying Brunauer-Emmett-Teller (BET) equation to nitrogen adsorption isotherms, which were measured by a surface area analyzer based on the volumetric method (BEL-Sorp max, BEL Japan, Inc.). The cross-sectional area of nitrogen molecule was taken as 0.162 nm^2. Existence of pore in the silica shell was investigated by t-plot method, using nonporous silica as the standard reference absorbent for the determination of pore sizes.

3. Results and discussion

Figure 1 shows SEM and TEM micrographs of a) PAA-NH_4OH template and b) hollow silica nanoparticles, which were prepared according to Yu's group [13]. The template shapes were totally spherical and have wide particle size distribution from several tens of nm to several hundreds of nm (Fig. 1a)). The shape and size distribution of the hollow silica nanoparticles (Fig. 1b)) were similar to that of the template. It can be said that the hollow silica nanoparticles formed as a replica of the PAA-NH_4OH template.

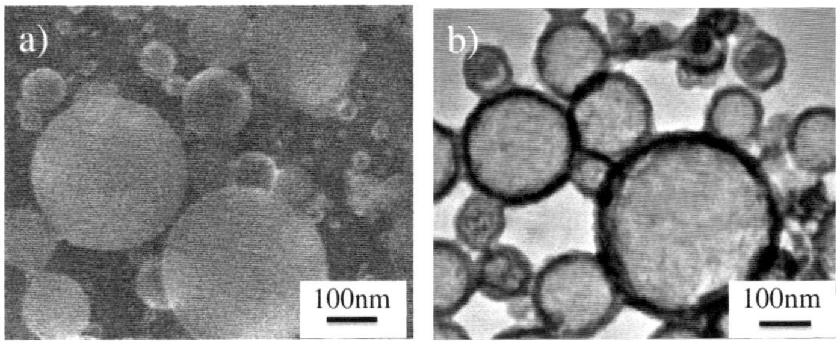

Figure 1. Microscopic observations of a) polyelectrolyte template and b) hollow silica particles by SEM and TEM, respectively.

Figure 2 shows change in particle size of PAA-NH_4OH template as a function of different NH_4OH amounts from 1.5 ml to 6.0 ml with PAA amount kept constant. In less than 1.5 ml NH_4OH, as-received PAA powder didn't completely dissolve in the NH_4OH solution. The amount of NH_4OH over 1.5 ml seems to be enough for preparation of PAA-NH_4OH mixture. The size of PAA-NH_4OH template significantly decreased with increase in NH_4OH amount. For 1.5, 3.0, 4.0, 5.0, 6.0 ml NH_4OH amounts, synthesis of hollow particles were attempted in the same way as described above.

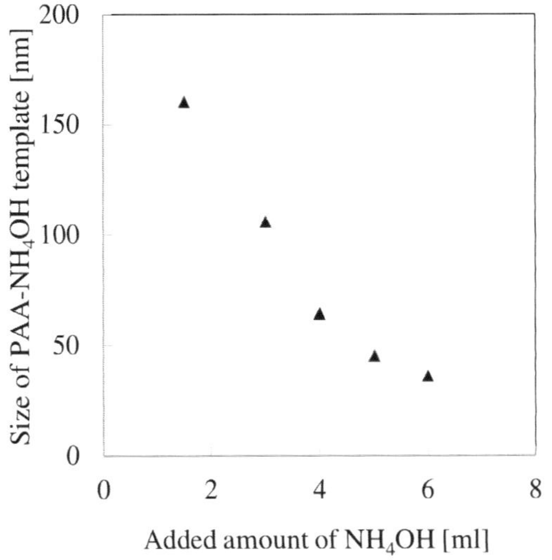

Figure 2. Particle size of PAA-NH_4OH template as a function of NH_4OH amount.

Figure 3 shows STEM images of obtained samples with NH_4OH additions of a) 1.5, b) 3.0, c) 6.0 ml, respectively. In the 1.5 ml NH_4OH addition, hollow particles were obtained (Fig. 3a)). They were spherical particles with wide particle size distribution. While in the range of 3.0 and 6.0 ml additions, only dense nanoparticles were observed without hollow silica nanoparticles as shown in Fig. 3b) and c). When PAA-NH_4OH mixture was dropped in EtOH, insoluble PAA-NH_4OH mixture forms spherical droplet in EtOH. In the droplet, PAA molecules aggregate each other owing to their intra- and inter-molecular interactions through hydrogen bond, ionic interaction between negative PAA and positive NH_4^+ ion. When TEOS was added in the PAA-NH_4OH/EtOH suspension, only hollow silica particles were obtained with the NH_4OH of 1.5 ml without dense particle formation (Fig. 3a)). Therefore, it can be said that NH_4^+ ion in the droplet works as a catalyst for sol-gel reaction of TEOS and the hydrolyzed TEOS reacts at only the PAA-NH_4OH droplet surface. An increase in NH_4OH in the droplet means increases in both NH_4^+ ion and water in the droplet. The increasing water in the droplet makes the droplet difficult to maintain the shape in EtOH, because effect of the interactions becomes low. To keep the droplet shape, the excess water in the droplet will leak from the droplet to outer phase EtOH. This could be a reason why that particle size decreased with increase in NH_4OH as shown in Fig. 2. Along with the water leakage, polarity of EtOH increases. The increasing NH4+ ions in the droplet also leak to EtOH, because higher polarity is favourable for NH_4^+. The NH_4^+ ions in the outer phase EtOH work as sol-gel catalyst in advance of that in the droplet. When TEOS added in the suspension, the hydrolyzation and condensation of TEOS easily occur in EtOH using the leaked NH_4^+ ion as a catalyst before TEOS approach to the droplet surface. Therefore, dense silica nanoparticles were formed with excess amount of NH_4OH, which is over than 3.0 ml in this experiment.

Figure 3. STEM images of obtained samples with NH_4OH additions of a) 1.5, b) 3.0, c) 6.0 ml.

From these results, if the conditions were optimum, TEOS reacts at only the PAA-NH_4OH droplet surface and only hollow silica nanoparticles were obtained. To design silica shell microstructure of the hollow nanoparticles, control of sol-gel reaction rate at the droplet surface is required. According to our previous results, sol-gel conditions like pH and reaction time control silica shell formation, e.g., coarse silica shell forms at low pH and for short time reaction and vice versa [12]. Here, polarity of EtOH where the PAA-NH_4OH droplets were dispersed was focused. Effect of water addition in the outer phase EtOH on microstructure of silica shell was investigated with 1.5 ml of NH_4OH kept constant.

Figure 4 shows STEM images of obtained samples prepared a) without water addition, with b) 0.1, c) 1.0, d) 2.0 ml water additions. In Fig. 4a)b)c), hollow silica particles were observed and their shell morphologies were changed with water amounts. Without water addition (Fig. 4a)), the silica shell was relative thick and rough silica shell surface was observed. With 0.1 ml addition (Fig. 4b)), the silica shell thickness decreased and the shell surface roughness also slightly decreased. In these hollow particles without water and with 0.1 ml, fine silica particles which composed of silica shell were recognized. On the other hands, with water addition of 1.0 ml (Fig. 4c)), the shell thickness significantly decreased and smoothness of the silica shell increased compared to the former two hollow particles. The fine silica particles were not recognized in the silica shell and it seems to have the highest density silica shell in the three. With further water addition up to 2.0 ml (Fig. 4d)), hollow particles were not formed and only dense particles were observed. When PAA-NH_4OH mixture was dropped in EtOH, the PAA-NH_4OH droplets form "soft" PAA aggregation by NH_4^+ ions linkage including water. In the droplets, these molecules move freely during stirring, therefore NH_4^+ ions which appeared at the droplet surface by chance can work as sol-gel catalyst when TEOS approaches to the droplet surface. This is the reason why this technique needs long time to form silica shell coating at the droplet surface. When water added in the outer phase EtOH, polarity of EtOH increases because of water's high polarity. Because TEOS is rather hydrophobic before hydrolyzation owing to their alkoxide groups, TEOS approaching rate to the droplet surface depends on the reaction solvent polarity. In other words, when EtOH polarity increases by water addition, TEOS approaching rate becomes slowly. By comparing Fig. 4a) (without water) with Fig. 4c) (with 1.0 ml addition), TEOS approaching rate can be faster in the former and slower in the latter. In the latter, for the small amount of TEOS by slower TEOS approaching, there were rich NH_4^+ catalyst in/at the droplet, which results in dense and thin silica shell formation (Fig. 4c)). When the excess amount of water added in EtOH, the PAA-NH_4OH droplet could be unstable because PAA-NH_4OH droplet dissolves in water. In the droplet, PAA molecular movement becomes more active and silica shell formation at such unstable droplet surface must be difficult. Therefore, only dense particles were formed as shown in Fig. 4d).

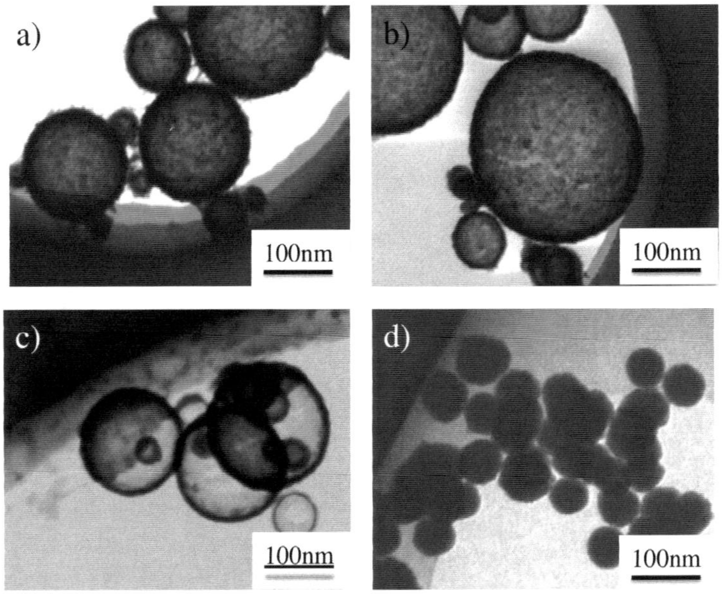

Figure 4. STEM images of the samples prepared (a) without water addition, with addition of (b) 0.1, (c) 1.0, (d) 2.0ml water in EtOH.

Shell microstructures of the hollow particles without water addition, with 0.1, 1.0 ml addition were investigated. Figure 5 shows the nitrogen gas adsorption/desorption isotherms of them (left) and a representative *t*-plot (right). The specific surface areas of them were calculated as 181, 71, 469 m^2/g by BET equation, and pore sizes of them were calculated as 2.2, 2.0, 1.3 nm by *t*-plot. The hollow particles, synthesized without and with 0.1 ml water addition, have mesopores in their silica shells. While the hollow particles synthesized with 1.0 ml water addition have micropores in the silica shell. It indicates possibility that water addition in EtOH controls silica shell microstructure of hollow particles.

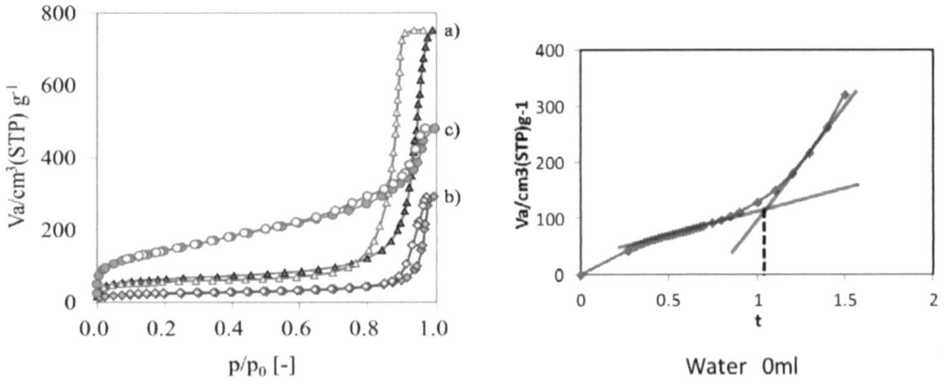

Figure 5. N_2 adsorption/desorption and a representative *t*-plot of the obtained hollow silica particles (a) without and with (b) 0.1, (c) 1.0 ml water in EtOH.

4. Conclusion

The controllable synthesis and shell structure design of hollow SiO_2 nanoparticles were achieved by optimizing the NH_4OH and water amount in reaction solution. In the case of NH_4OH amount below a critical value, hollow SiO_2 nanoparticles could be obtained. Furthermore, the shells formed in water-free solution were thick, rough and rich of mesopores, which became relatively thinner, smoother and richer of micropores if the water amount was gradually increased within a limited range.

Acknowledgement

This work was supported by the Advanced Low Carbon Technology Research and Development Program (ALCA) of the Japan Science and Technology Agency (JST).

References

[1] Fielding L A, Mykhaylyk O O, Schmid A, Pontoni D, Armes S P and Fowler P W 2014 *Chem. Mater.* **26** 1270
[2] Yang J, Lee J, Kang J, Lee K, Suh J S, Yoon H G, Huh Y M and Haam S 2008 *Langmuir* **24** 3417
[3] Guerrero-Martínez A, Pérez-Juste J and Liz-Marzán L M 2010 *Adv. Mater.* **22** 1182
[4] Caruso F, Caruso R A and Möhwald H 1998 *Science* **282** 1111
[5] Fuji M, Shin T, Watanabe H and Takei T 2012 *Adv. Powder Technol.* **23** 562
[6] Virtudazo R V R, Tanaka H, Watanabe H, Fuji M and Shirai T 2011 *J. Mater. Chem.* **21** 18205
[7] Caruso R A, Susha A and Caruso A 2001 *Chem. Mater.* **13** 400
[8] Li T, Moon J, Morrone A A, Mecholsky J J, Talham D R and Adair J H 1999 *Langmuir* **15** 4328
[9] Fowler C E, Khushalani D, Mann S 2001 *Chem. Comm.* **20** 28
[10] Li Y, Shi J, Hua Z, Chen H, Ruan M and Yan D 2003 *Nano. Lett.* **3** 609
[11] Tan B and Rankin S E 2005 *Langmuir* **21** 8180
[12] Takai C, Watanabe H, Asai T and Fuji M 2012 *Colloids Surf.* **404** 101
[13] Wan Y and Yu S-H 2008 *J. Phys. Chem. C* **112** 3641

Elaboration and characterization of metallurgical silicon for photovoltaic applications

M Barbouche, M Hajji, F Krout and H Ezzaouia

Laboratoire de Photovoltaïque, Centre de Recherches et des Technologies de l'Energie, Technopole de Borj-Cédria, BP 95 Hammam-Lif 2050, Tunisia.

E-mail: barbouchemed@gmail.com

Abstract. There is a small quantity of participants in the global market of silicon, mainly from the developed countries. It should be noticed also that production of metallurgical silicon Mg-Si is among the most important steps to produce solar grade silicon and photovoltaic panels. Therefore, in this paper we focused on the growth of Mg-Si by carbothermal reduction of silica. An investigation was made using FT-IR characterization to study the effect of process conditions (temperature, atmosphere, duration) in Mg-Si production. Raman spectroscopy was used to investigate the produced Mg-Si. Based on these results, we established a pilot line production of metallurgical silicon at the "CRTEn" in Tunisia.

1. Introduction

Photovoltaic (PV) technology is expected to be a leading technology to solve the issues concerning the energy and the global environment that we are now facing, due to several advantages of the PV system [1]. Silicon is the most popular material for solar cells. Metallurgical silicon is produced by the reduction of silica in an electric arc furnace by means of carbonaceous materials. In the submerged arc furnace, silica and carbon raw materials react under atmospheric pressure over a range temperature of 1573 K to 2273 K (1300 °C to 2000 °C).

The upper low temperature part of the furnace is called the outer zone, whereas the hottest part of the furnace the inner zone. There is general agreement about the reactions taking place in the inner zone of the furnace.

The main reactions in the inner zone are as follows [2]:

SiO_2 [s] + C [s] = SiO [g] + 2 CO [g] (1)

SiO [g] + 2 C [s] = SiC [s] + CO [g] (2)

SiC [s] + SiO [g] = 2 Si [s] + CO [g] (3)

2. Experimental

2.1 Sample preparation

The purity of silica used in this work is to 99.98% [3]. Common impurities in quartz, like Na, K, Al, and Fe come from minerals like mica and feldspar [4]. The carbon powder used is taken from coal. Its quantity was adjusted to obtain a nominal molar ratio C/SiO_2 = 3/1. Powders were mixed for more homogeneity.

Carbothermal reduction of the silica was performed in an electric arc furnace. The mixture loaded in the graphite crucible was placed into the hot zone of the furnace. The gas is introduced into the reaction chamber during the experiment whereas switching on the extraction system to let the exhausted gas flow into the ambient air. The temperature regulation is performed using temperature controller "OMRON E5AK"

2.2 Characterization

Synthesized samples were characterized by Fourier transform infrared spectroscopy (FTIR). The samples were ground and dispersed in a matrix of KBr at room temperature, the mixture was pressed into pellets. The spectra were saved in transmittance mode. Spectral resolution is 4 cm^{-1}.
Prepared samples were identified also by a Raman spectrometer (LabRAM HR JOBIN YVON Technology HORIBA Scientific) to determine SiC structure.

3. Results and discussion

3.1 FTIR analyses

The FTIR spectrum of powder synthesized under air, at T=1500°C and during 120s is shown in figure 1.

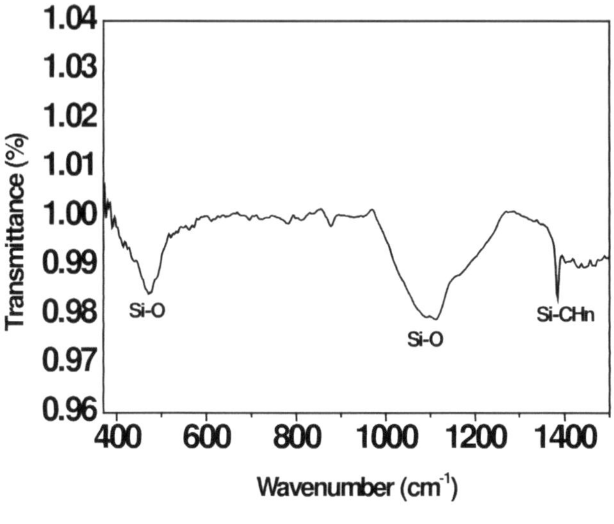

Figure 1. FTIR spectra of samples synthesized under air, at T=1500°C and during 120s.

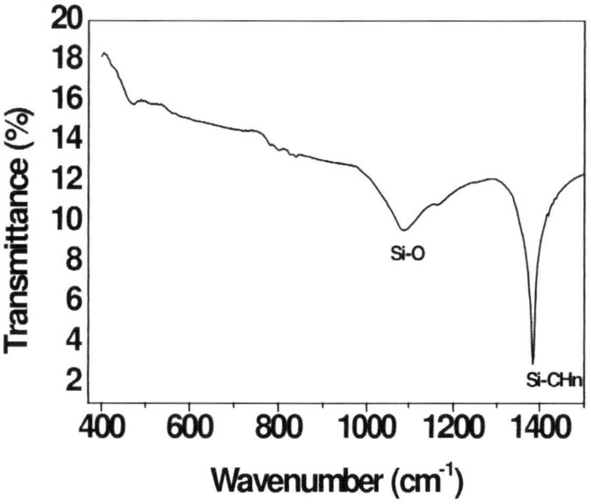

Figure 2. FTIR spectra of samples synthesized under air, at T=1700°C and during 120s.

The FTIR spectra given in Figure 1 shows three vibrations; Two peaks around 480 cm^{-1} and 1100 cm^{-1} attributed to Si-O [5] and a peak around 1385 cm^{-1} attributed to Si-CHn bond [6] which can be explained by the chemical etching realized before characterization. We increased temperature to 1700°C and we obtained FTIR spectra shown in figure 2. We observed decrease of Si-O peak intensity but we have no peak of Si-Si bond which means that we have a partial deoxydation of silica.

We changed the atmosphere to work under inert gas. Figure 3 illustrates the IR spectrum of the produced samples under argon, at T=1700°C and during 120s.

We notice the appearance of new peak around 810 cm^{-1} attributed to Si-C bond but we have no peak of Si-Si bond which means that we have also a partial deoxydation of silica.
We changed the atmosphere to work under CO and we increase the duration of treatment to 180s. We obtained FTIR spectrum illustrated in figure 4.

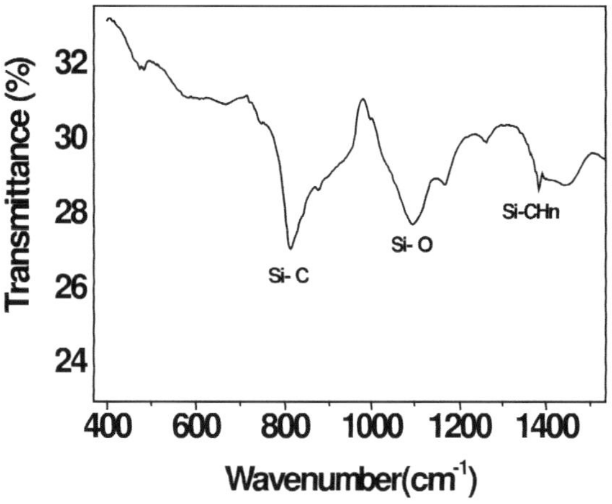

Figure 3. FTIR spectra of samples synthesized under argon, at T=1700°C and during 120s.

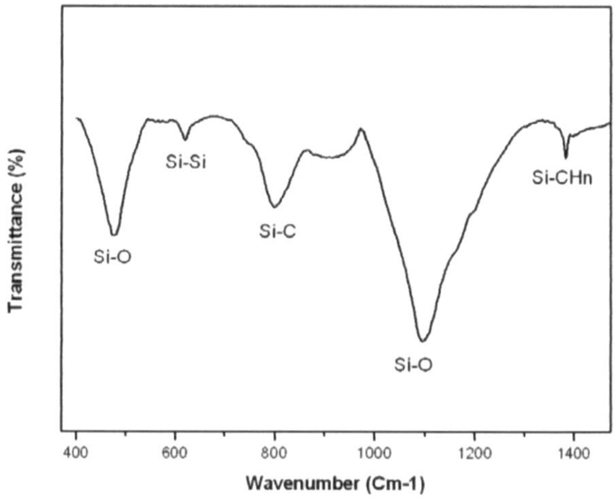

Figure 4. FTIR spectra of samples synthesized under CO at T=1700°C and during 180s.

As revealed in figure 4, a new peak around υ = 629 cm^{-1} appears. It's attributed to Si-Si bond. The peak around 480 cm-1 corresponds to Si-O bond, it can be explained by rinsing the sample in pure water before characterization. On the other hand the peak associated with Si-C bond remains, which means that we obtained a mixture of peak Si+SiC.

We increase again the temperature to 1900°C and we decrease the duration of treatment to 120s and we obtain the FTIR spectra illustrated in figure 5.

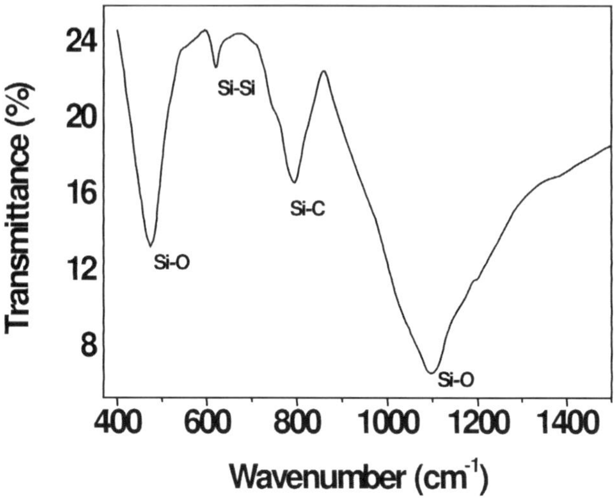

Figure 5. FTIR spectra of samples synthesized under CO, at T=1900°C and during 120s.

As shown in figure 5, we observe almost the same peaks of figure 4. These peaks are attributed to Si-O, Si-Si and Si-C bonds. This led us to increase the time of treatment to 180s by leaving temperature at 1900°C and under CO, we obtain then FTIR spectra shown in figure 6.

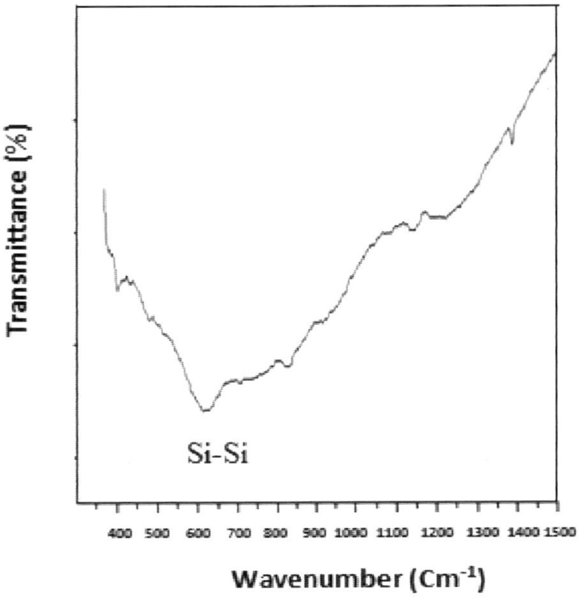

Figure 6. FTIR spectra of samples synthesized under CO at T=1900°C and during 180s.

IR analysis (figure 6) shows that the major peak is around υ = 629 cm^{-1} due to vibrations of the Si-Si bond. This means that we obtain silicon when we synthesize our samples under CO, at T=1900°C and during 180s.

3.2 Raman analyses

Figure 7 shows Raman spectra of used as raw materials in our work.

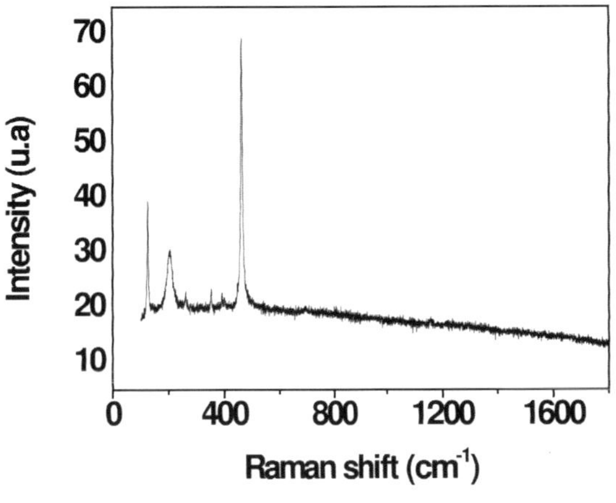

Figure 7. Raman spectra of SiO_2.

Figure 8 illustrates the Raman spectrum of the produced samples under CO, at T=1900°C and during 180s.

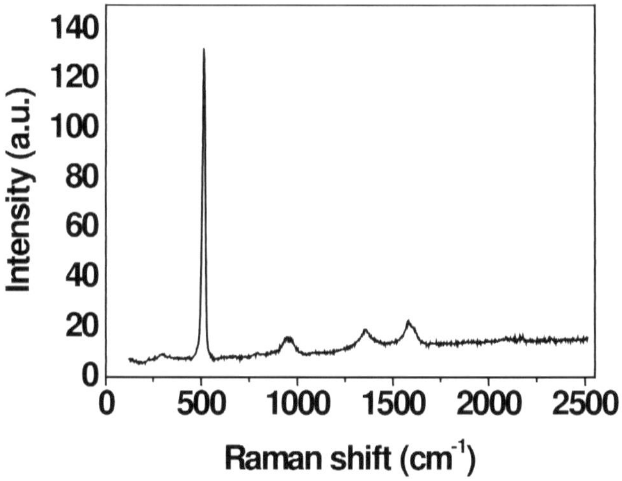

Figure 8. Raman spectra of samples synthesized under CO, at T=1900°C and during 180s.

The analysis of this spectrum shows the existence of a single major peak at 517 cm^{-1}. This peak is attributed to the Si-Si bond [7]. This means that we have succeeded in producing silicon when we synthesize our samples under CO, at T=1900°C and during 180s which confirms results obtained by FTIR analyses. The cleanliness of the peak tells us that our silicon is crystallized. The other three peaks are less important and are attributed to Si-C and CC.

4. Conclusion

As a conclusion, this work allowed finding the conditions (electric power, atmosphere and duration of growth) that promote the growth of metallurgical silicon. This work enabled to set up a knowledge base for the manufacturing technology of metallurgical silicon from silica and carbon. Metallurgical silicon is well produced in Tunisia. The next step is to purify the Mg-Si to produce solar grade silicon.

References

[1] Kato K, Murata A, Sakuta K 1997 *Sol. Energ. Mat. and Sol. Cells* **47** 95-100
[2] Dal Martello E, Tranell G, Gaal S, Raaness O S, Tang K, and Arnberg L 2011 *Metallurg. and Mat. Trans. b* **42B** 939-950
[3] Khalifa M, Hajji M, Ezzaouia H 2012 *EDP Science*
[4] Dal Martello E, Tranell G, Gaal S, Arnberg L 2011 *Metallurg. and Mat. Trans.* **42** 939-950
[5] Prabhakaran P V 2009 *J. of Mat. Sci.* **44** 528–533

[6] Janz S 2006, Master in Physics. Fraunhofer Institute for Solar Energy Systems Freiburg 22
[7] Goldberg B, Swan A K 2007 *Raman measurements of Stress and strain in Silicon* Boston University

Journal of Physics: Conference Series **596** (2015) 012009 doi:10.1088/1742-6596/596/1/012009

Morphology Changing at Incipient Crystallization Condition

Takeshi Toshima[†1], **Ryo Hamai**[2], **Saya Fujita**[1], **Yuka Takemura**[1], **Saori Takamatsu**[1], **Masamoto Tafu**[1]

[1]National Institute of Technology, Toyama College. 13 Hongo-machi, Toyama, 939-8630, Japan
[2]Kyushu Institute of Technology, 2-4 Hibikino, Wakamatsu, Kita-Kyushu, 808-0196, Japan

E-mail[†]: t.toshima@nc-toyama.ac.jp

Abstract. Brushite (Dicalcium phosphate dihydrate, (DCPD), $CaHPO_4•2H_2O$) is one of key components in calcium phosphate system due to wide attractive material not only as bioceramics but also environmental materials. Morphology of DCPD crystals is important factor when one uses its functionality with chemical reaction; because its surface crystal face, shape and size rule the chemical reactivity, responsiveness. Moreover, physical properties are also changed the morphology; such as cohesion, dispersiveness, permeability and so on. If one uses DCPD crystals as environmental renovation materials to catch the fluoride ions, their shape require 020 crystal surfaces; which usually restricts their shape as plate-like structure. After the chemical reaction, the shape of sludge is not good for handling due to their agglutinate property. Therefore searching an effective parameter and developing the method to control the morphology of DCPD crystals is required. In past, we reported that initial concentration and pH value of starting solution, prepared by dissolving calcium nitrate, $Ca(NO_3)_2$ and ammonium dihydrogen phosphate, $NH_4H_2PO_4$, changes the morphology of DCPD crystals and phase diagram of morphology of DCPD crystal depend on those parameter. The DCPD crystallization shows unique behaviour; products obtained higher initial concentration form single crystal-like structure and under lower condition, they form agglomerate crystal-like structure. These results contradict usual crystallization. Here we report that the effect of mixing process of two solutions. The morphology of DCPD crystals is changed from plate structure to petal structure by the arrangement. Our result suggests that morphology of DCPD crystals strongly depends at incipient crystallization condition and growth form is controllable by setting initial crystallization condition.

1. Introduction

DCPD (dicalcium phosphate dihydrate, $CaHPO_4 \cdot 2H_2O$) crystal is one of key components in calcium phosphate system and used as a coating for orthopaedic implants, starting materials of other calcium phosphate [1-2]. In recent year, DCPD crystal is researched as an environmental treatment material for removing fluoride ions and heavy metal ions from waste water and soil by transition to FAp (fluoroapatite, $Ca_{10}(PO_4)_6F_2$) and partly heavy metal substituted-FAp [3].

On the other hands, controlling the shape of such the functional materials attracts material scientist due to increasing their potential as following reasons; their size and surface condition (specific surface area and ratio of the crystal face with good reactivity) rule chemical reactivity and/or responsibility and physical properties. Additionally, after the reaction with target ions, size and shape of sludge sometimes changes their usability. In previous works, plate shaped crystals show good reactivity with fluoride ions; nano-sized precursor often appears at 020 crystal surface. Under aqueous solution synthesis without impurity, DCPD crystals often form plate structure with large 020 crystal surface. But this shaped crystals easily stack each other and without good dynamical property such as mobility and diffusibility. Especially, after the waste water treatment, shape of FAp would change the transit cost of sludge due to its moisture content. Therefore DCPD morphology changing into spherical shaped and located with 020 crystal surface is required.

In past report, we tried to controlling the morphology of DCPD crystals by adding other ions during aqueous synthesis [4], setting the initial pH value and concentrations of calcium ion and phosphate ion [5]. And we succeeded to change the morphology of DCPD crystals from plate-like structure to petal-like structure by preceding parameters. But the results shows paradoxical with usual crystal growth; under crystal growth condition with higher concentration, morphology of DCPD crystals becomes to single crystal-like plate structure and under lower concentration, morphology changes to multi-crystals; aggregation of plate structure (petal). In typical crystallography, under higher concentration condition, many crystal nuclei appear everywhere and they aggregate with each other. Therefore, morphology changing of DCPD crystals shows unique phenomenon.

Here we report to unveil the mechanism of morphology changing of DCPD crystals by very simple approach; by changing the mixing process of two solutions. We designed the synthesis process; initial crystallization condition was set under the condition with petal-like structure and crystal growth condition was set under the condition with plate-like structure. Solid phase products were observed by scanning electron microscope (SEM). Mixing process designing changed the morphology of final products.

2. Experimental Method

2.1. Strategy for Unveiling mechanism of Morphology changing

In crystallography, morphology changing mechanisms is classified two types; one is depended on the shape of crystal nuclei; crystallization condition rules the morphology and shape of crystal nuclei is kept during crystal growing. And the other one is depended on crystal growing condition; morphology of crystals changed due to anisotropic crystal growth by the difference of crystal growth speed like the Berg effect [6] or difference of stability of crystal face [7].

Therefore, we designed synthesis and judge which mechanisms change the morphology of DCPD crystals. Figure 1 is phase diagram of morphology of DCPD crystal depended on initial pH value and concentration; which is modified from the past report [5]. We started the synthesis at the condition of "no product" and passed the area of "petal-like product", finally, reached to the condition of "plate-like product". And we kept the total-environment of crystal growth with same value with last condition. If morphology of DCPD crystals is ruled by initial condition, shape of final products change to petal-like structure, and it is ruled by the environment of crystal growing, shape of final products must be same results with past report [5].

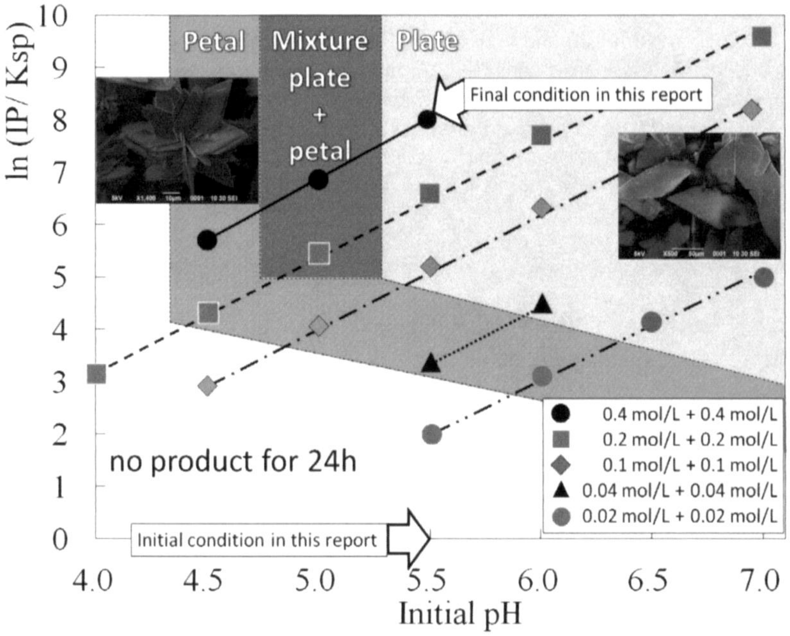

Figure 1. Phase diagram of morphology of DCPD crystal depends on initial pH value and super-saturation ratio [5]. Arrowed condition (pH=5.5 and ln (IP/KsP)=0) is starting condition and other arrowed condition (pH=5.5 and ln (IP/KsP)≒8) is final condition of this report.

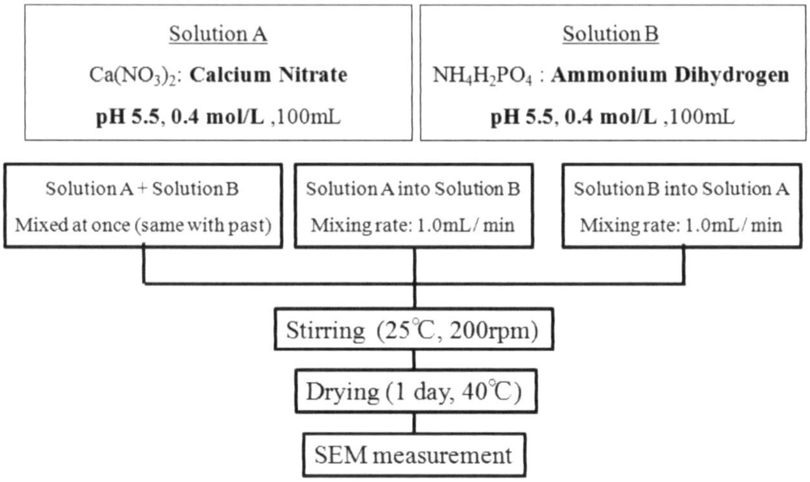

Figure 2. Experimental flow diagram of this report. We synthesized DCPD crystals with three-kind processes.

2.2. Experimental flowchart

DCPD crystals were synthesized according to the flow diagram shown in Figure 2 under aqueous condition by mixing two solutions; solution A (calcium nitrate ($Ca(NO3)2$, Wako, 98.5%)) and solution B (ammonium dihydrogen phosphate ($NH4H2PO4$, Nacalai, 99%)), were prepared by dissolving preceding chemicals and adjusted initial concentration and initial pH value by the addition of 0.01 and 0.1 M nitric acid (Kanto Chemical) and ammonia solution (Wako, 25%). In this report, we set the initial pH value is 5.5 and initial concentration is 0.4 mol/L; in past report, under this condition, all of products form only plate-like structure. Solution A (100 mL) and solution B (100 mL) is mixed (1) at once with stirring (200 rpm) and kept 60 min, this is same process with past report; (2) solution A is dropped into solution B with stirring, the mixing ratio is 1.0 mL /min and stirring speed is same with (1), finish the dropping, we kept stirring 60 min and (3) solution B is dropped into solution A with same process with (2). Obtained solid-products were filtered through a 0.45 μm membrane filter. The solid products were dried at 40 °C for 24 h. The products from each process were observed by SEM (JSM-6390AX, JEOL, accelerating voltage of 3 kV without coating).

Figure 3. SEM images of each mixing process. (A) Mixed two solutions at once. (B) Calcium ion solution into phosphate ion solution. (C) Phosphate ion solution into calcium ion solution.

3. Experimental Results and Discussions

Figure 3 (A), (B) and (C) are SEM images of products obtained under different mixing processes with magnification 200 and 2000. 3-(A) is the result with process-(1); two solutions were mixed at once and all product form simple plate-like structure, they are same with past report [5]. On the other hands, 3-(B) and 3-(C) are results with process-(2) and (3), respectively. In the case of process-(2), morphology of products is basically formed with plate-like structure, but few petal-like structured products are also obtained. And in the case of process-(3), most products form petal-like structure and few plate-like structured products are obtained. Our results show the morphology of DCPD crystal is changed only mixing process; this is equal to say, DCPD morphology changing is depended on former mechanism. Incipient crystallization condition changes the shapes of crystal nuclei and they are kept during crystal growing process. At the beginning, crystallization condition is on the area of "no product for 24h" in figure 1. About 5 minutes later from the starting the dropping, crystallization condition would be into the area of "petal-like structure". And after 7 minute from the starting, concentration of ion would finish the crossing area of "petal-like structure" and be into the area of "plate-like structure". These times are estimated on simple assumption; each ion would not combine and form DCPD crystals. Actually, staying time on the area of "petal-like structure" would be longer due to DCPD crystallization (But final crystal growth condition is same with (1)). Moreover, shape of one leaf of petal is basically same with plate-like products. Crystal surface is 020 and anisotropic crystal growth direction is same with all mixing process. Therefore, during crystal growing process, shape of product does not change. We also compare the lower initial concentration (0.2 mol/L and the other experimental flow is same), the rate of petal-like structure reduces but same tendency is also obtained.

4. Summary

We synthesized DCPD crystals under aqueous condition and arranged the mixing process of two starting solutions and obtained following results.

(1) Morphology of DCPD crystals is changed from plate structure to mixture condition (plate and petal structure) by changing the mixing process.

We compare the shape of products on mixing process dependency (each starting solutions are set that initial pH value is 5.5 and initial concentration is 0.4 mol/L or 0.2 mol/L) (a) at once and (b) one solution into the other with dropping rate 1.0 ml/min.

(2) Morphology of DCPD crystals is kept during crystal growing and ruled crystal nucleation condition.

Our results suggest that morphology of DCPD crystals is controllable by only setting incipient crystallization condition (and then growth larger with keeping crystal nuclei shape). In future work, we would like to evaluate chemical and physical properties of these petal DCPD crystals.

ACKNOWLEDGEMENTS

This work was supported by Japan Society for the Promotion of Science (JSPS) KAKENHI Grant Number 23310058.

References

[1] Kumar M, Dasarathy H and Riley C, 1999 *J. Biomed. Mater. Res.* **45,** 302
[2] Furutaka K, Monma H, Okura T and Takahashi S. 2006 *J. Eur. Ceram. Soc.* **26**, 543
[3] Tafu M and Chohji T 2006 *J. Eur. Ceram. Soc.* **26**, 767
[4] Hamai R, Toshima T, Tafu M, Masutani T, Chohji T, 2013 *Key Engineering Materials*, **529-530**, 55
[5] Toshima T, Hamai R, Tafu M, Takemura Y, Fujita S, Chohji T, Tanda S, Li S, Qin G W, 2014 *J. Asian Ceramic Soc.* **2**. 52
[6] Berg W F, 1938 *Proc. R. Soc. London.* **A164** 79
[7] Kuroda T, Irisawa T, Ookawa A 1977 *J. Cryst. Growth.* **42** 41

A control strategy for PV stand-alone applications

S Slouma[1, 2] and H Baccar[1]

[1] Technopôle Borj Cédria, Centre de Recherche et de Technologie de l'Energie, LR10CRTEn01 Laboratoire Photovoltaïque, 2050, Tunisie.
[2] Université de Tunis El Manar, Ecole Nationale d'Ingénieurs de Tunis, LR11ES15 Laboratoire de Systèmes Electriques, 1002, Tunis, Tunisie.

E-mail: safa_slouma@yahoo.com

Abstract. This paper proposes a stand-alone photovoltaic (PV) system study in domestic applications. Because of the decrease in power of photovoltaic module as a consequence of changes in solar radiation and temperature which affect the photovoltaic module performance, the design and control of DC-DC buck converter was proposed for providing power to the load from a photovoltaic source.In fact, the control of this converter is carried out with integrated MPPT (Maximum Power Point Tracking) algorithm which ensures a maximum energy generated by the PV arrays. Moreover, the output stage is composed by a battery energy storage system, dc-ac inverter, LCL filter which enables higher efficiency, low distortion ac waveforms and low leakage currents. The control strategy adopted is cascade control composed by two regulation loops.Simulations performed with PSIM software were able to validate the control system.The realization and testing of the photovoltaic system were achieved in the Photovoltaic laboratory of the Centre for Research and Energy Technologies at the Technopark Borj Cedria. Experimental results verify the effeciency of the proposed system.

1. Introduction

A photovoltaic system (PVs) converts light energy into electricity. The basic device of a PVs is the PV cell. PV Cells are grouped to form modules or Panels. Modules may be grouped to form large PV arrays. A PV cell is basically a p-n semiconductor junction; the use of monocrystalline solar cell is recommended in many PV applications on account of its efficiency photovoltaic solar cells. In fact, Monocrystalline solar panels have the highest efficiency rates since they are made out of the highest-grade silicon. The efficiency rates of monocrystalline solar panels are typically 15-20% live the longest. The performance of a PV system depends also on the atmospheric conditions. The PV power, output voltage and current vary as functions of temperature and solar radiation. Hence, the I–V and the P–V characteristics of the PV system are nonlinear, and there is an optimum operating point (MPP) on the characteristics that can supply maximum power to the load. Therefore, Perturbation and Observation control algorithm (P&O) is proposed to extract the Maximum Power Point (MPP) at all times, irrespective of the type of PV panel or weather conditions [1, 2].

An inverter is inevitable at the output of the first stage, in order to generate a single-phase-level. In this stage, a LCL filter is necessary to reduce the total output signal harmonic distortion.

Because of the intermittent PV production, this system needs to be secured by storage system such as battery bank at the output of the first stage. Another reason for battery is the system applications context which is specific to isolated locations or countryside.

2. Proposed Stand-Alone PV System

The proposed System depicted in Figure 1 is composed by dc-dc converter connected to PV system, tracking the maximum power point of the panels. Connected at the output of the first stage, the inverter generates an output sinusoidal waveform. In this stage, a LCL filter was performed to reduce the total output signal harmonic distortion. Furthermore, a battery bank is connected at the output of the first stage.

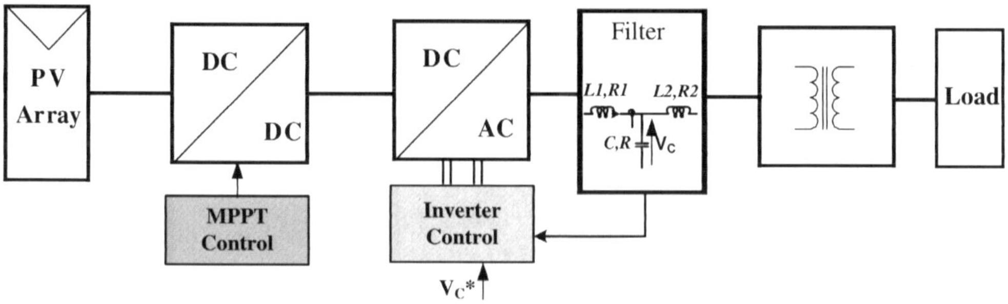

Figure 1. Stand-Alone PV System.

3. Experimental Setup

3.1. System Overview

The test bench presented in Figure 2 is installed in the Centre for Research and Energy Technologies at the Technopark Borj Cedria. It allows to determine the electrical characteristics of the solar panel.
The experimental system is divided of PV array connected to an acquisition chain and computer. The PV is Eurosolar M 510 A, whose electrical specifications are presented in Table 1. The radiation is measured using a pyranometer sensor. Also, the array surface temperature is measured using a thermocouple. Then, the experimental data was recorded for different atmospheric conditions and the result graphs were plotted.

PV panel simulated is Eurosolar M 510 A. The PV parameters are summarizing in Table 1. PV system is composed by 8 PV panels of 50 W, as shown in Figure 3, resulting in 400 W of total installed power. PV system is further interconnected in a 4 PV parallel-series configuration. Initially, each PV panel is producing 50 W at a1000 W/m² radiation

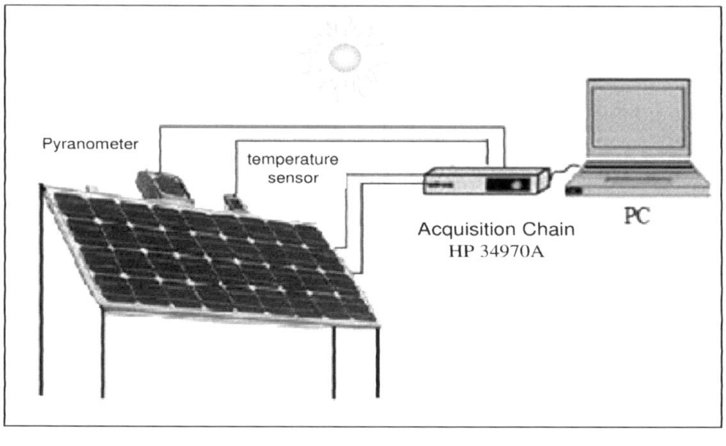

Figure 2. Test bench Photovoltaic system.

Figure 3. Photo-voltaic system.

Table1. PV panel Electrical STC specifications.

Panel Type	Eurosolar M 510 A
Maximum Power	50 W
Optimal Current	2.95 A
Optimal Voltage	17 V
Short circuit current	3.25 A
Open circuit Voltage	21.4 V

3.2. Experimental results

Figures 4 and 5 prove that The PV current and power output vary with atmospheric parameters such as temperature and radiation. By increasing temperature, PV voltage will be drop. The PV Array current and Power decrease with decreasing radiation.

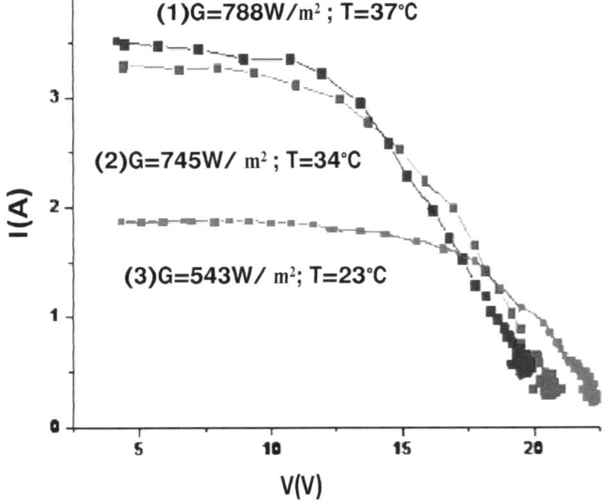

Figure 4. Characteristics I (V) curves of different temperatures and radiation.

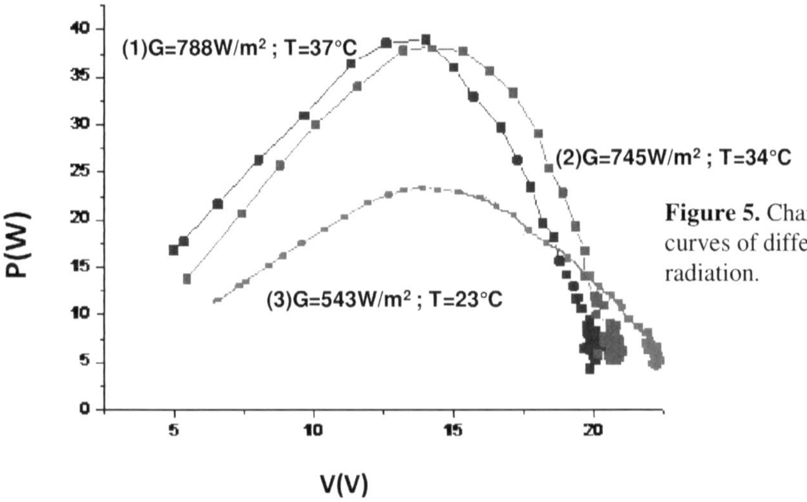

Figure 5. Characteristics P (V) curves of different temperatures and radiation.

The performance of the experimental system can be evaluated by comparing the measured array powers obtained from the experimental system with the theoretical one. In fact, theoretical results of the PV system are obtained from the PSIM-simulation software.

Figure 6 and 7 exposes the effect of radiation variations on the V-I and V-P PV characteristics. It is further seen that the PV current increases with increasing radiation (Figure 6). It is further observed that the PV power output increases with raising radiation (Figure 7). So theoretical and experimental results are consistent.

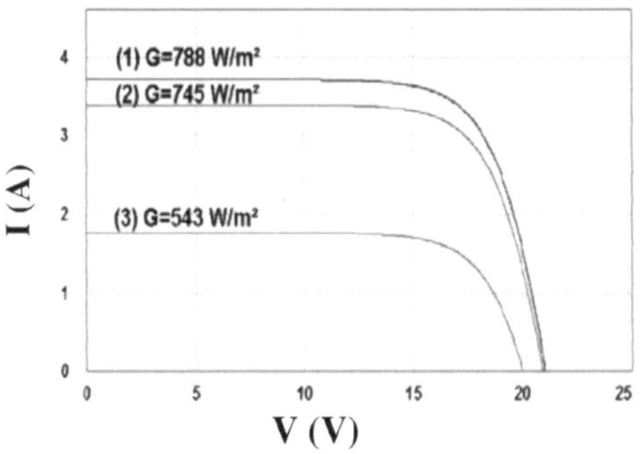

Figure 6. Characteristics I (V) curves of different radiations

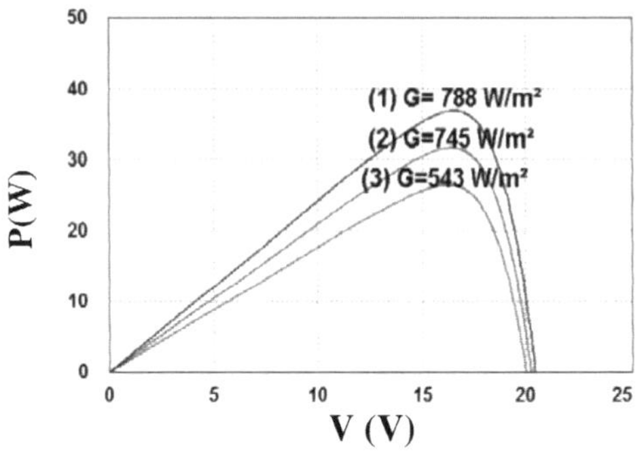

Figure 7. Characteristics P (V) curves of different radiations

4. Stand-Alone PV Control System

4.1. Control of single-phase PV-DC/DC-Inverter

4.1.1. Control of the DC-DC converter. Figure 6 and 7 show the V-I and V-P curves of a PV array, which each curve presents a maximum point at the MPP.
In order to guarantee that the maximum power will be extracted from the PV panels, a maximum power point tracker (MPPT) algorithm is required. Several algorithms are proposed in literature. Therefore, the Perturb & Observe (P&O) algorithm is widely used for its facility of implementation . Figure 8 belows the flow chart of the operation of the P&O MPPT algorithm.

After one perturb operation the current power is calculated and compared with previous value to determine the change of power ΔP. If ΔP>0, then the operation continues in the same direction of perturbation. Otherwise it reverses the perturbation direction. Figure 9 illustrates the MPPT simulation results where the PV power (PMPPT) follows its reference (Pref) at a1000 W/m² radiation [1], [2].

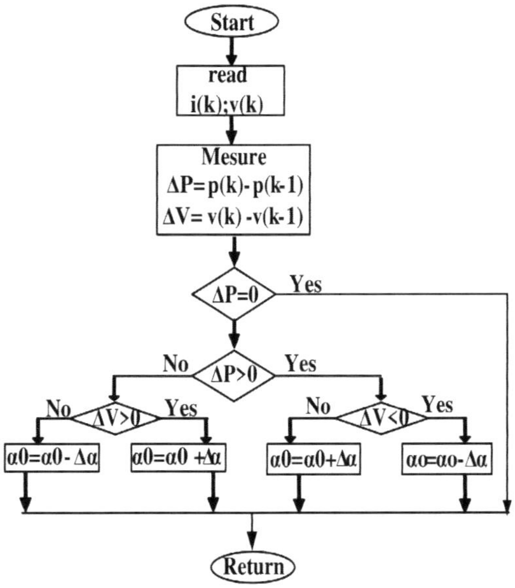

Figure 8. P&O MPPT flow chart.

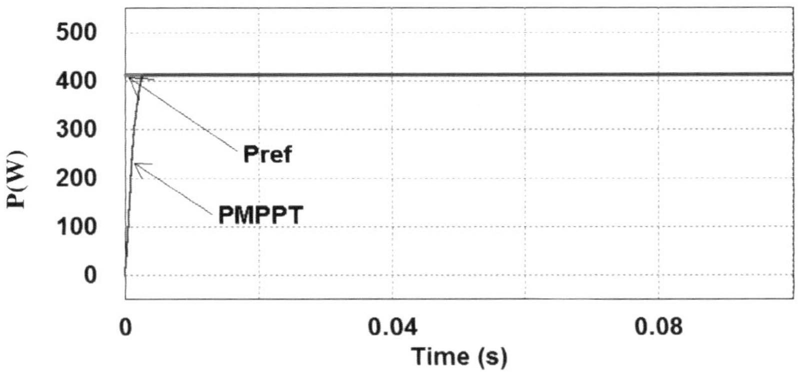

Figure 9. MPPT PV Power.

4.1.2. Control of the DC-AC inverter

4.1.2.1. Modeling filter. For the PV inverter in stand-alone system, the most common control structure for the dc-ac inverter is using a P+Resonant (PR) controller with a low-pass output filter. In fact, LCL filter is used that being a higher order filter (3rd) leads to more compact design.
The drawback is that due to its own resonance frequency which can produce stability problems. So a special control design is required.
The resonant frequency of the LCL filter is shown in:

$$f_{res} = \frac{1}{2\pi}\sqrt{\frac{L_{o1}+L_{o2}}{L_{o1}.L_{o2}.C_o}} = 1130 \text{ Hz} \tag{1}$$

A method for designing both the controller and LCL filter ensuring stability is shown in Figure 10.
4.1.2.2. Synthesis of the closed loop control

The DC-AC Control inverter is performed where P+Resonant (PR) is introduced and applied to single phase PWM inverter control. In this approach the PI dc-compensator is transformed into an equivalent ac-compensator. The control strategy adopts a cascade control loop which is composed of two internal and external regulation loops. The cascade control loop of the PV inverter with PR controller is depicted in the Figure 10.

The P+Resonant (PR) controller is defined as: $C(s) = K_p + \frac{K_I}{s^2 + \omega_0^2}$ (2)

The PI controller H(s) is defined as: $H(s) = \frac{1}{L_{o1}s + R_{o1}}$ (3)

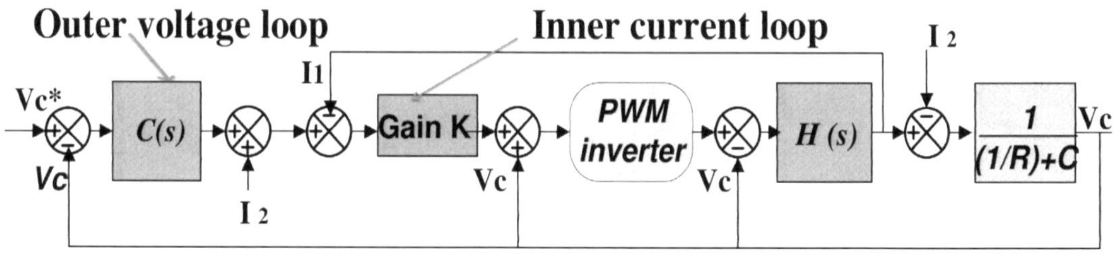

Figure 10. Closed control loop Synthesis.

Figure 11 and 12 expose the effect of LCL filter use. Indeed, in Figure 12 the total output signal harmonic distortion is reduced compared with the signal in figure 11.The issue of stability is becoming more important when LCL filter is used.

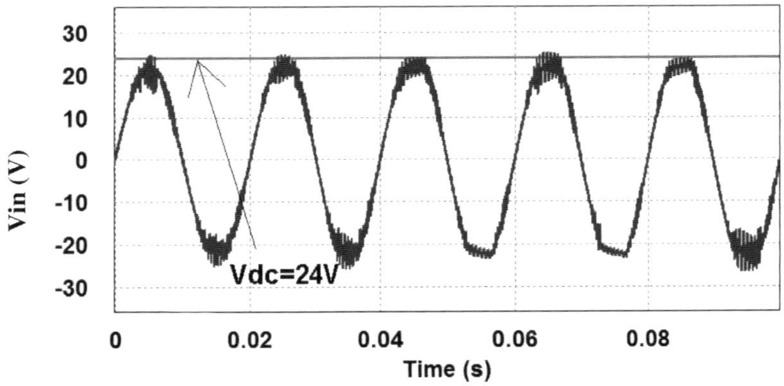

Figure 11. Voltage before Filter and Transformer.

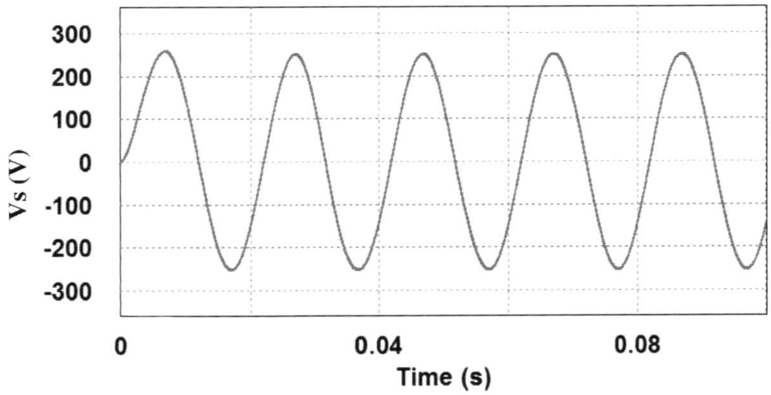

Figure 12. Voltage after Filter and Transformer.

5. CONCLUSIONS

This paper proposed a stand-alone PV system. The experimental results confirm the theoretical ones discussed above.

By means of simulation results, the efficiency of the proposed structure, as well as its control system, were verified. A method for designing both the controller and LCL filter ensuring stability is developed.

ACKNOWLEDGEMENTS

This work was supported by the Tunisian Ministry of High Education and Research under CRTEn-LR10CRTEn01.

REFERENCES

1. Narayanappa, Thanushkodi K 2012 *University "Politehnica" of Bucharest (U.P.B.) Sci. Bull., Series C* **74** 261
2. Manimekalai P, Harikumar R., Aiswarya R. 2012 *IJCA Proceedings on International Conference on Advances in Communication and Computing Technologies (ICACACT)* **3** 25-30

Chemical prosperity of various chemical gypsums from viewpoint of particle morphology

M Tafu[1], T Toshima[2], A Manaka[1], T Chohji[3], M Irie[4]

[1]Department of Applied Chemistry and Chemical Engineering, National Institute of Technology, Toyama College, Japan, 13 Hongo-machi, Toyama 939-8630 Japan

[2]Department of Mechanical Engineering, National Institute of Technology, Toyama College, Japan, 13 Hongo-machi, Toyama 939-8630 Japan

[3]President, National Institute of Technology, Kagoshima College, Japan, 1460-1 Shinkou Hayato-cho, Kirishima city 899-5193 Japan

[4]Faculty of Life and Environmental Sciences, University of Tsukuba, Tenodai 1-1-1, Tsukuba, Ibaraki, Japan

E-mail: tafu@nc-toyama.ac.jp

Abstract. Dissolution rate of various chemical gypsums in the water was investigated by using batch experiment. The rate of the dissolution of the gypsum had good agreement for explanation of dissolution of plate particle. The rate constants of the experimental formula were different with particle morphology of the gypsums. When morphology of the particle was sheet-liked, the dissolution rate was 10 timed large than it of block-shaped particle. From result of observation of dissolution phenomena of the in the water, it was find that dissolution of the gypsum was preceded on long axis. From these results, particle morphology is seems to be important parameter for using chemical gypsums.

1. Introduction

Chemical gypsum is calcium sulfate dihydare ($CaSO_4 \cdot H_2O$) produced from various chemical industries such as phosphate, fluoride and titanium production, and the fuel gas desulfurization (FGD) process in coal-fired power plant. Many amounts of these chemical gypsums are used for plasterboard in wall of the building. After demolition of the building, many amount of waste plasterboards are generated and disposed to landfill. In Japan, recycling of the waste plasterboard is strongly demanded because of lack of landfill capacity. By production of phosphate from phosphate rock, five tons of waste gypsum generated by production of one ton of phosphoric acid. Therefore, reuse of the gypsum (phosphogypsum) is one of the important issues for phosphate industries [1]. For another chemical industry, reuse of the byproduct gypsum is important on viewpoint of ecological and economical aspects.

Major application of recycling of the waste gypsum is ground stabilizer for soft ground and/or construction sludge [2, 3]. The application is considered that gypsum have hydraulic property and make hardening soil or sludge. This property is caused by "calcium sulfate". Gypsum has various particle shapes, and particle and chemical property may be different. It was reported that particle size of the gypsum is able to increase by addition of sodium sulfate [4], and sodium tripolyphosphate [5]. Variation of particle shape is affect to powder property such as fluidity. In case of gypsum, dissolution rate and adhesion property of water molecule are different by crystal orientation [6, 7].

In this study, dissolving property of various chemical gypsums was investigated. From results of the experiment, it was appeared that dissolving rate of the chemical gypsum was controlled by specific crystal face.

2. Experimental

2.1. Materials and chemicals

Chemical gypsums used in this study were byproduct from phosphate production and the FGD process. In comparison, synthesized gypsum from conventional precipitation process was also used. Two point five grams of the gypsum was mixed with 1 L of the pure water. The water purified by Mill-Q water purifier (Mill-Q Academic, Merck-Millipore).

Particle shape of the gypsums was observed by optical microscope (ECLIPSE LV100D, Nikon, Japan). Figure 1 shows particle shape of the chemical gypsums used in this study. FDG gypsums (samples A and B) were block-liked shape, and phosphogypsums were plate-liked shape.

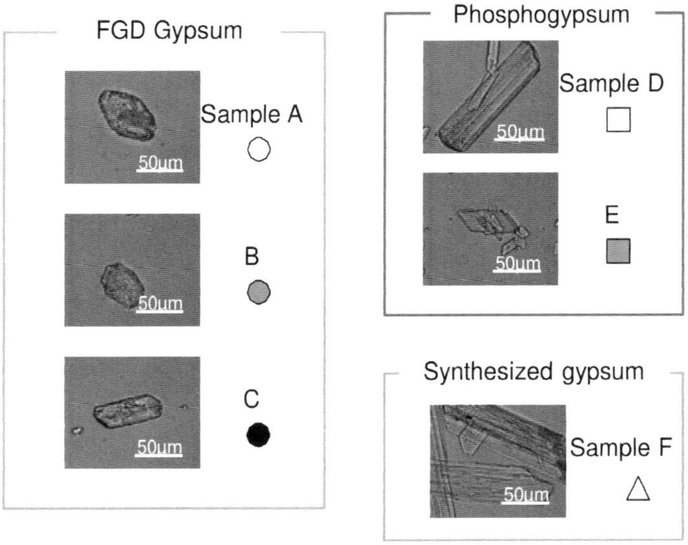

Figure 1. Particle shape of the chemical gypsums used in this study.

2.2. Dissolution of the gypsums

Mixture of water and the gypsum stirred by baffled mixer (MBF-250-ME, ELENA, Japan). Small amount (about 2-3 mL) of the mixture was sampled every 10 seconds. The sampled mixtures were filtered by 0.45 μm of membrane filter. Calcium and sulfur concentration were analyzed by inductively coupled plasma atomic emission spectrometer (ICP-AES, 720ES, Agilent-Varian).

2.3. Observation of dissolution of the gypsums

The optical microscope was used for observation dissolving phenomena of the gypsums in the water.

3. Results and discussions

3.1. Dissolution rate of the chemical gypsums

Dissolution rate of the various chemical gypsums in the water was investigated by using batch mixer. Mixture of the gypsum and water collected every 10 seconds and filtered. Calcium and sulfur concentrations were analyzed by ICP-AES. Figure 2 shows change in concentrations of calcium and sulfate (calculated by concentration of sulfur) ions in the water after added gypsum (sample A).

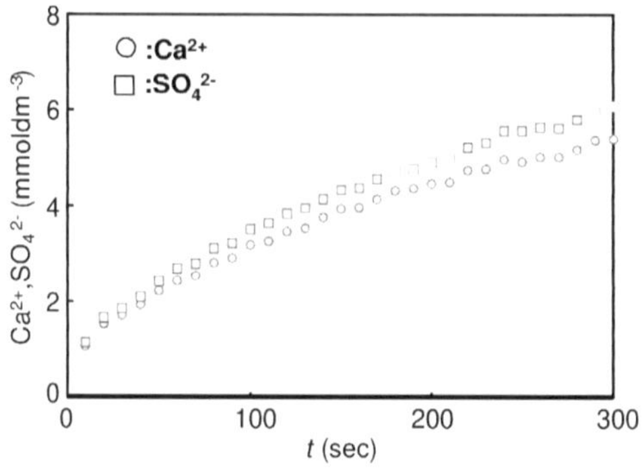

Figure 2. Change in concentration of calcium and sulfate ions in the water after added gypsum.

To characterize dissolving rate of the gypsum, model equation for dissolving plate-shaped particle in water was applied. The model equation is equation (1) [8].

$$kt = [1 - (1\text{-X})^{1/2}]^2 \tag{1}$$

X of the equation indicate fraction of the dissolving ions (X = 0 to 1), t is time and k is apparent rate constant. By using equation (1), result of Figure 2 was standardized to liner relationship like Figure 3. The k value was calculated by change in calcium concentration in the figure.

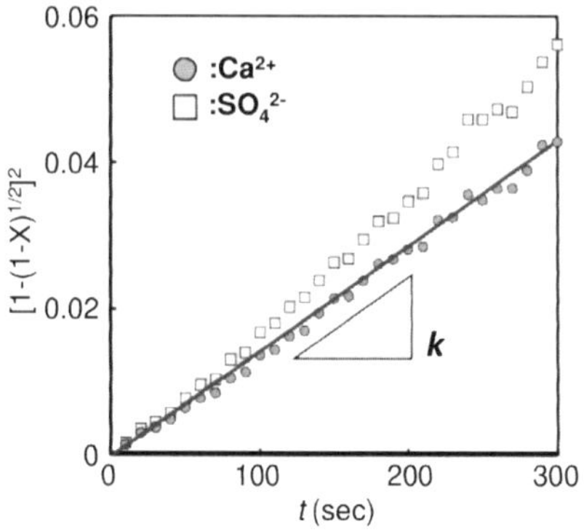

Figure 3. Standardized of change in concentration of by equation (1).

By using this standardization, result of samples were shown in Figure 4. The figure shows results of samples A, D and F. The figure shows dissolving rate of the gypsum was quite different of each samples. Figure 5 shows variation of the apparent rate constants of all samples. The figure also shows particle shape of the samples. From the figure, the k value was smaller in case of particle shape was block-liked.

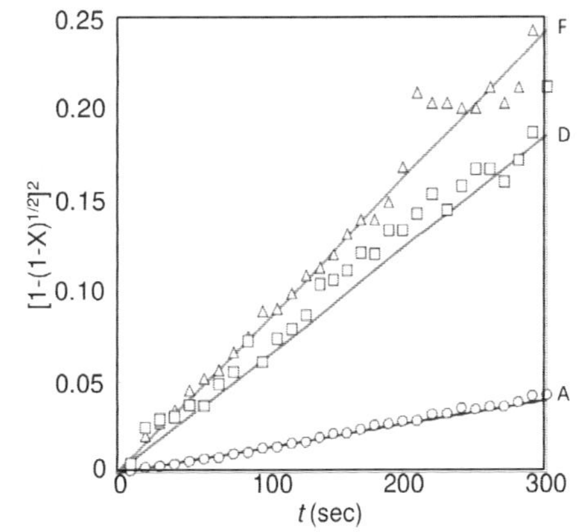

Figure 4. Dissolving rate calculation of all samples.

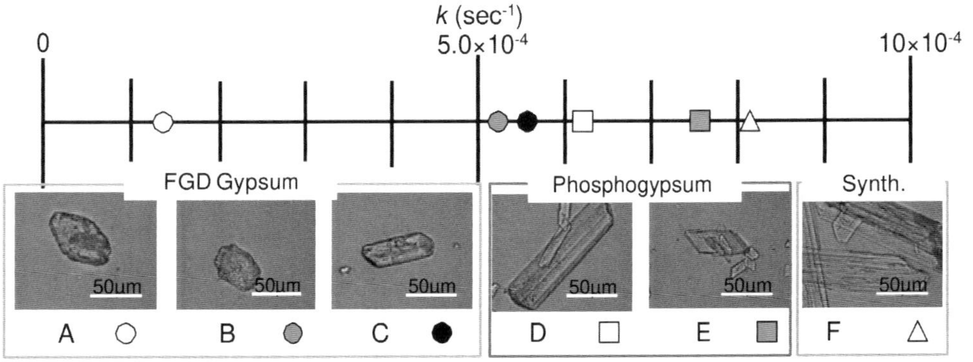

Figure 5. Variation of the *k* values.

3.2. Dissolution phenomena of the chemical gypsums

To discuss difference of the *k* values in the chemical gypsums, dissolution phenomena of the gypsums in the water were observed by using optical microscope. Change of shape of the particle was observed every 5 minutes. Sample A and F were selected because of difference of the *k* value. Figure 6 shows change of particle shape after immersion of the gypsums into water.

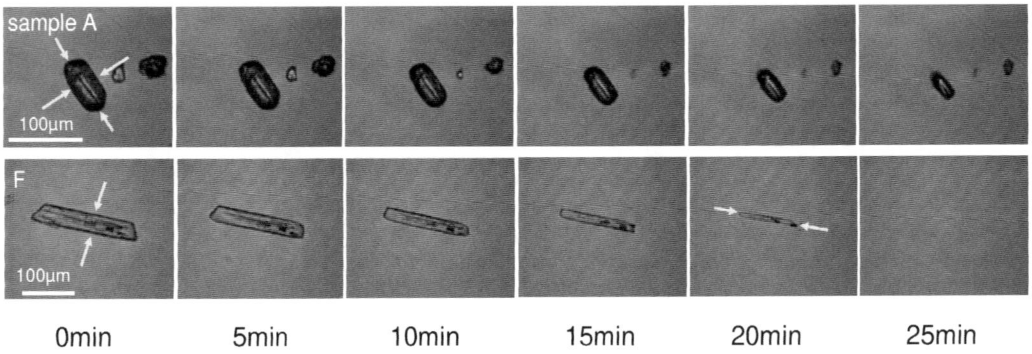

Figure 6. Dissolution of gypsum particle.

Dissolving rate of the sample F was larger than it of sample A. In case of the sample F, dissolving was started on minor axis and particle shape changed to slim. After that, dissolving was observed on major axis of the gypsum particle. On the other hand, dissolving of sample A was isotropic.

Particle morphology of sample F was seems to be typical shape of gypsum crystal. Flat face of the particle was (010). Dissolving rate of (010) face is smaller than another face such as (110) and (111) [9]. From this results, orientation of crystal is seems to be predominant to dissolving rate.

4. Summary

In this study, dissolution rate of various chemical gypsums in the water was investigated by using batch experiment. Obtained results were as follows;

1) The rate of the dissolution of the gypsum had good agreement for explanation of dissolution of plate particle.
2) The rate constants of the experimental formula were different with particle morphology of the gypsums.
3) When morphology of the particle was sheet-liked, the dissolution rate was 10 timed large than it of block-shaped particle.
4) From result of observation of dissolution phenomena of the in the water, it was find that dissolution of the gypsum was preceded on long axis. Especially, crystal orientation of (010) face is one of the important factors of controlling dissolving rate.

From these results, particle morphology is seems to be important parameter for using chemical gypsums.

Gypsum is used another applications. Tsunami disaster on Higashi-Nippon earthquake made salt-affected soil in large area. Addition of gypsum is one of the solutions to remediate the salt-affected soil by ion exchange between sodium and calcium ions from the gypsum [10]. In case of usage of gypsum for remediation of the salt-affected soil, supply of calcium ion from gypsum is one of the important properties. For usage of the chemical gypsum for various applications, understanding and controlling of particle shape of the chemical gypsum is seems to be important aspects.

Acknowledgement

The authors deeply appreciate the helpful support with the experiments given by Dr. Hiroyuki Oda (He is now in Hosokawa Micron Group). A part of the study was supported by research grant from Gypsum Board Association of Japan.

References

[1] Villalba G, Liu Y, Schroder H and Ayres RU 2008 J Ind. Ecology **12** 557
[2] Tafu M and Chohji T 2008 *J. Soc. Inorg. Mater. Japan* **21** 421
[3] Tafu M, Chohji T, Fujita T, Hiwasa M and Nakano H 2008 Japanese Patent No. 4748608
[4] Kojima Y and Yasue T 2006 *J. Eur. Ceram. Soc.* **26** 777
[5] Yu Q and S Veeramasuneni 2002 US Patent 6409825
[6] Finot E, Lesniewska E, Mutin JC and Fundonnet JP 1997 *Sur. Sci.* **384** 201
[7] Karshin VP and Grogorja VA1970 *Zh. Fiz. Khim.* **44** 1356
[8] Kim EY, Lee JC, Kim BS, Kim MS and Jeong J 2007 *Hydrometallurgy* **86** 89
[9] Arai Y and Yasue T 1989 *Powder Sci. Eng.* **21** 62
[10] Sakai Y, Ren S, Wang C and Sadakata M 2011 *J. Chem. Eng. Jpn.* **44** 750

Optical properties of nanostructured TiO_2 thin films

M Ben Karoui[1], Z Kaddachi[2] and R Gharbi[2]

[1]. Laboratoire de Photovoltaïque, Centre de Recherche et des Technologies de l'Energie, Technopole de Borej-Cedria, BP 95, Hammam-Lif, Tunis. 2050, Tunisia.

[2]. LSDE-C3S. Ecole Nationale Supérieure d'Ingénieurs de Tunis. Université de Tunis, 05 Av. Taha Hussein 1008 Montfleury. Tunis, Tunisia.

E-mail: moufida_gharbi@yahoo.fr

Abstract. Nanostructured titanium dioxide (TiO_2) thin films were prepared with one and two layers on glass substrates (ITO) using sol gel-spin coating technique and annealed at temperature ranging from 300°C to 600°C. X-ray diffraction (XRD) and UV-Vis transmittance spectroscopy were used to study the effects of annealing temperature on the structural and optical properties. X-ray diffraction studies showed that the as-deposited films were amorphous and at first changed to anatase phase with increase of annealing temperature. Optical constants of these films were derived from the transmission and reflection spectra subjected to annealing at different temperatures. The results indicate that an anatase phase structure TiO_2 thin film with nanocrystallite size of about 10.72 nm can be observed at 400°C. The optical gap decreases from 3.82 eV to 3.74.eV for one layer and from 3.64 eV to 3.5 eV for two layers when the annealing temperature increases from 300 °C to 600°C. The refractive index is low compared to a pore free anatase TiO_2 and it increases contrary to the porosity which is reduced with the raising of the annealing temperature.

1. Introduction

TiO_2 thin films have specific characteristics such as the chemical stability, the non-toxicity, a high refractive index and a good transmittance in the visible region associated with a strong absorption in the ultraviolet. These properties allow their use in many applications: Optical components [1], ultraviolet attenuating agents for sunscreen cosmetic applications [2], optoelectronic sensors, photocatalyst [3,4]. They are also a good candidate for the fabrication of the dye sensitized solar cells DSSC [5].

Titanuim dioxcide is an excellent semi-conductor which crystallizes in three different crystalline structures: anatase (tetragonal), rutile (tetragonal) and brookite (orthorhombic). At room temperature, the direct gap energy E_g is 3.06 eV for rutile and 3.3 eV for anatase. These two values of the band gap allow a high absorption in the ultraviolet range. The anatase is formed at lower temperatures and sharing some properties with rutile such as hardness and density. At high temperature, around 800°C it will turn into rutile [6]. TiO_2 thin films have been prepared by many techniques such as reactive sputtering, sol gel [7], and chemical vapor deposition [8]. However, the sol gel technique attracts more attention by its low processing coast, simplicity and ability to produce a uniform thin film with a good control in the preparation conditions. TiO_2 thin films are obtained using molecular precursors via inorganic polymerization reaction with titanuim alkoxcides [9]. The optical properties of the TiO_2 anatase films prepared by sol gel method deposed by spin coating are studied.

2. Experimental details

Nanocrystalline TiO_2 thin films are prepared using the sol-gel spin coating techniques. The sol was obtained by the partial hydrolysis and condensation of tetrabutyl orthotitanate $Ti(OC_4\,H_9)_4$ with H_2O, the butanol and acetic acid used as solvent and catalyst. The procedure of preparation includes the dissolution of 1 mol titanuim butoxide as precursor, 1 mol of H_2O, 4 moles of butanol and 3 moles of acetic acid. After mixing the adequate proportions for 90 min, we obtain at the final a very transparent solution ready for the deposition. The gel was spin coated on the ITO substrate at 2500 rpm for 30 s for one and two coating. After, the film was dried at 100°C for 15 min and then heat treated at different temperatures ranging between 300°C and 600°C with increasing temperature rate of 5°C min $^{-1}$ for 2 h in a furnace.

The thin films have been systematically characterized after preparation and their eventual heat treatment. X-ray diffraction was used as a first characterization step to identify the structures of the crystallized phases present in the thin films. The diffractometer used was a Siemens D5005 Bragg–Brentano geometry (40KV, 20A) equipped with a copper anticathode wiring with CuKα radiation (λ= 1.5406 Å). The diffractograms were recorded at room temperature and the X-ray data were collected from 20° to 80° with a measurement time of 10 s. UV–visible spectroscopy studies were performed in order to determine the optical properties for wavelengths ranging from 250 to 2500 nm using a UV–visible-near infrared spectrometer PerkinElmer (Lambda-950) equipped with an integrating sphere.

3. Results and discussion

3.1. Structural properties

The XRD characterization of the annealed thin films obtained after one and two coating, were shown in figure 1(a, b). It can be seen that the diffraction pattern of TiO_2 annealed at 300°C does not exhibit a clear peaks and especially the peak around 25.35° is absent which indicates that the films are amorphous. However, from 400°C the intensity of the peaks around 25.35° was improved and they well detected, the diffraction pattern shows another small peak with a weak intensity characteristic of the anatase phase. At 500°C and 600°C, the peaks characteristic of anatase phase centered around 25.35° were defined revealing the nucleation and growth of grains. Using the JCPDS data, the peaks are well defined, there are 6 diffraction peaks showed at 25.35°, 37.56°, 41.45°, 44.48°, 45.45°, 55.72° and assigned (101), (004), (210), (200), (111), (211) planes respectively and correspond to anatase phase characterized by unit cell parameters (a=b=3.78) and c = 9.516. Such a type of behavior has also been noticed by A.Ranjitha et al. [10] who used sol gel-dip coating for the deposition, they obtain anatase phase with high intensity peaks at 500°C.

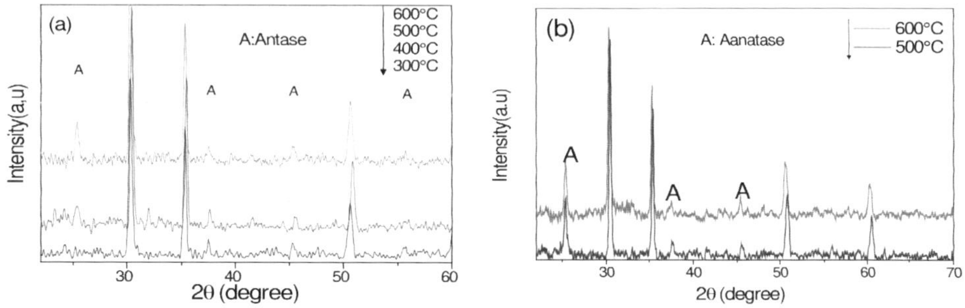

Figure 1. Evolution of diffraction patterns of the TiO_2 thin films after (a) one coating and (b) two coating for various annealing temperature. A=antase.

As we raise the temperature of annealing, we observe an increase of the intensity correspond to the peaks which are centered around 25.24° characteristic of anatase phase. This late revealed an improvement of the crystalline structure. Also, it can be seen that the intensity of the peaks is very higher for films having two layers than one layer. This result is may be due not only to the improvement of the crystalline quality but also to the increase in proportion of titanium oxide and the density of layer. There is no peak corresponding to rutile phase which agrees with the literature showing that the films change their crystallinity nature from anatase to rutile when they annealed below 600°C [11]. The crystallite size of TiO_2 thin films can be calculated from XRD patterns using Scherrer equation [12].

$$L = \frac{K\lambda}{\beta \cos\theta} \tag{1}$$

Where L is the crystallite size of TiO_2 thin films, K is a constant equal to 0.94, λ is the wavelength of X-ray (CuK_α= 1.5406 Å) and β is the true half –peak width (FWHM). Table 1 shows the variation of the grains size for different annealing temperature and number of layers. As the temperature increases, TiO_2 crystallites continue to develop and increase from 25.85 nm to 31.30 nm for two layers and from 10.72 nm to 20.97 nm for one layer. This results show that there was an improvement of the connections of Ti-O-Ti bond of the anatase phase as a function of the annealing temperature and the number of layer.

3.2. Optical characterizations

Figure 2 shows evolution of the optical transmittance spectra of the thin films of TiO_2 (one, two layers) for various annealing temperature from 300°C to 600°C. All spectrums are characterized by the presence of two different regions: The first region of high transparency showing a few interference oscillations which are due to multiple reflections at the film/substrate and the film/air interfaces. The second region corresponds to strong absorptions characterized by a sharp drop in optical transmission. This abrupt drop, which is the fundamental absorption of thin layers, is due to the electronic transition between the valence and conduction bands of semiconductor materials.

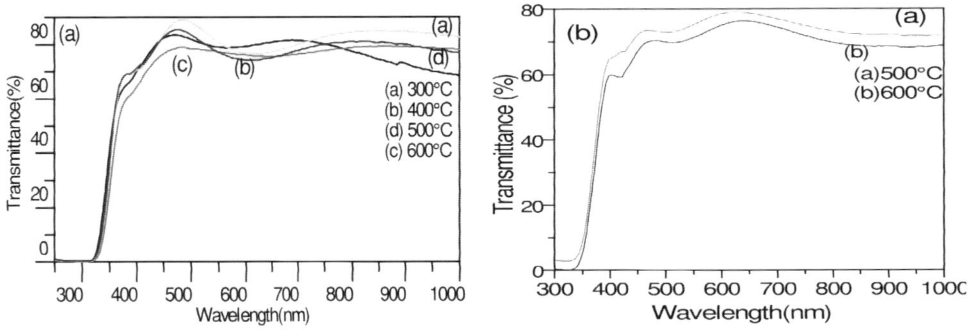

Figure 2. Optical transmittance spectra of TiO_2 thin films at different annealing temperatures: (a) one layer, (b) two layers.

All the films obtained after one coating exhibit a transmittance higher than 80% in the visible region; they are highly transparent in the visible and opaque in the ultraviolet region. This high transparency is one of the properties that explain the interest of TiO_2 thin films in solar cells applications. We find that the transmittance percentage of TiO_2 thin films decreased slightly during annealing from 300°C to 600°C. This is due to the increase of the light diffusion with the crystallite size and particle aggregation. Thin films with one layer exhibit a higher transmittance than two layers. The increase in numbers of layers leads to a decrease in the transmittance percentage. The reflectance of all films reaches over 20% in the visible region and it is clear that the thin film having one layer exhibits a reflectance higher than two layers.

Tablel. Refractive index, porosity and thickness of TiO_2 thin films for different annealing temperatures.

	Eg (eV)		**Porosity (%)**		**Refractive index**		**Thickness (nm)**		**L (nm)**	
Temperature (°C)	One layer	Two layers	One layer	Two layers	One layer	Two layers	One layer	Two layers	One Layer	Two layers
300	3.92	-	40.03	-	2.05	-	70.36	-	8.84	-
400	3.87	-	30.28	-	2.17	-	65.69	-	10.72	-
500	3.80	3.64	28.82	42.40	2.19	2.02	60.67	95.40	18.09	25.85
600	3.74	3.50	15.01	32.68	2.35	2.14	61.66	94.67	20.97	31.30

The optical band gap energy was estimated using the Tauc's model [13]. Table 1 shows the energy gap values of TiO_2 thin films annealed at various temperatures. We observe that the gap energy of one layer decreases from 3.88 eV to 3.74 eV and from 3.64 eV to 3.5 eV for two layers with increasing annealing temperature. This decrease is due to the structural properties of the thin films. In fact, the annealing reduces the internal stresses and randomness formed during deposition and it affects the adjustment of relative position of atoms. The improvement in the crystallinity of the films and the decrease in the density of the grain boundaries with increasing the temperature affect very well the energy gap, this behavior has mentioned also by Ranjitha et al. [10].

3.3 *Methods for calculating refractive index of thin films.*

We used Heavens model to determine the thickness and the optical absorption energy gap coefficient of thin films, from analysis of their spectrum transmission $T(\lambda)$ and reflectivity $R(\lambda)$. This model is particularly used in the field of low absorptions. To determine the thickness of the thin film, we must first calculate the refractive index n_c. The expression n_c of a weakly absorbent thin layer deposited on a transparent substrate was calculated from the measured UV-Vis reflectance spectrum in the region of a weakly absorbing film [14]:

$$n_c = \sqrt{n_s \times \frac{(1+\sqrt{R_{\max}})}{(1-\sqrt{R_{\max}})}} \tag{2}$$

Where n_s is the refractive index of the substrate and R_{max} is the maximum average reflectance layer. The thickness d of the layer can then be deduced from the following relation:

$$d = \frac{m \times \lambda}{4 \times n_c} \tag{3}$$

Where m is the order of interference and λ is the wavelength. The porosity P of the thin films is calculated using the following equation [15]:

$$P = \left(1 - \frac{n_c^2 - 1}{n_d^2 - 1}\right) \times 100(\%) \tag{4}$$

Where n_d is the refractive index of pore–free anatase (2.52) and n_c is the refractive index of the thin film. From table 1, it is clear that the refractive index of TiO_2 thin films increases with enhancing temperature treatment, contrary to the porosity which decreases. This is interpreted by the increase in the size of the grains and densification of the films and pore destruction in films during the increasing temperature treatment. In consequence the particles are electronically interconnected [16].

We see that variation of thin layers thickness depends on the temperature and it can be correlated with structural properties of thin film. Indeed, increasing temperature, the film becomes more crystallized and the grain size increases, therefore his compactness was enhanced and the film thickness decreases. The same observation was mentioned by Michiakh et al. [17]. Refractive index $n(\lambda)$ of the layers as a function of wavelength λ was obtained from the transmission spectra in the weakly absorbing region of these films. We use the method of envelopes, it consists to draw two envelopes corresponding to the extrema of the interference fringes (an envelope for the maxima and minima for another transmission) and to take a value on each of these envelopes for the same wavelength λ rated T_{max} and T_{min} respectively. The refractive index $n(\lambda)$ was given by:

$$n(\lambda) = \sqrt{s + \sqrt{s^2 - n_0^2(\lambda) n_s^2(\lambda)}} \tag{5}$$

Where n_0 is the refractive index of air, n_s is the refractive index of substrate and s is determined by this expression:

$$s = \frac{1}{2(1+n_s^2)} + 2n \times \frac{T_{max}(\lambda) - T_{min}(\lambda)}{T_{max}(\lambda) \times T_{min}(\lambda)} \tag{6}$$

Figure 3 shows the spectral variation of the refractive indices $n(\lambda)$ of the thin films (one and two layers) of TiO_2 for 500°C and 600°C. We note that for the low wavelengths, $n(\lambda)$ reaches a maximum for a value λ_m and it shifts to longer wavelengths by increasing the annealing temperature. In the area of transparency, $n(\lambda)$ stabilizes and tends to be constant. This maximum λ_m is interpreted as the energy corresponding to the optical gap of the thin layer. The optical gap energy of the layers decreases with the annealing temperature therefore $n(\lambda)$ increases from 2.09 to 2.11 and from 2.17 to 2.18 for one and two layers respectively.

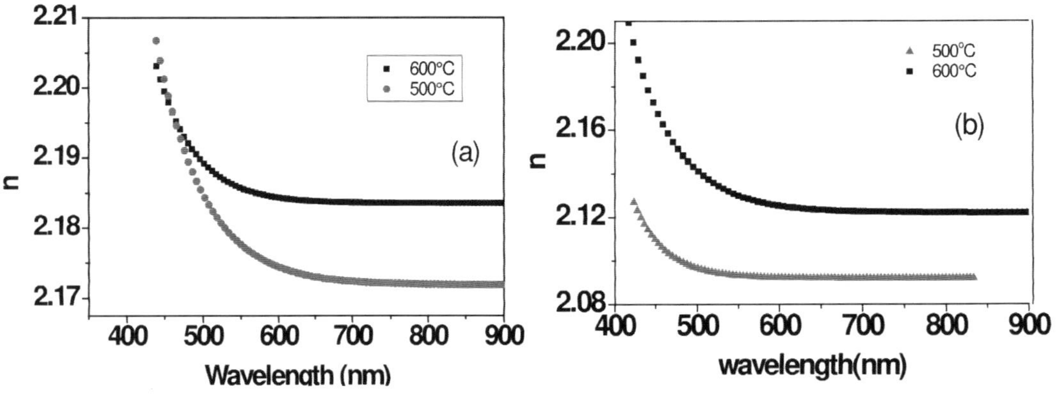

Figure 3. Refractive index as a function of wavelength of TiO_2 films: (a) two layers, (b) one layer.

Present values of refractive index are slightly lower than the reported values in anatase single crystal (2.50). It could be due to partial crystallinity and low atom mobility of the films at room temperature. It has been noticed that the refractive index and band gap energy show opposite trends with annealing. Such a type of behavior has also been reported by Khan et al. [18].

4. Conclusion

We have prepared nanocrystalline TiO_2 thin films deposited on ITO substrates using tetrabutyl-orthotitanate precursor by employing a simple and inexpensive sol–gel spin coating technique. We have observed that films crystallize in the anatase phase after annealing at 400°C and the grain size showed depends on annealing temperature indicating recrystallization. Band gap of the film decreased from 3.92 eV to 3.74 eV for one layer and from 3.65 eV to 3.5 eV for two layers when the temperature varied from 300°C to 600°C. The as deposited films were porous and were found to become dense as annealing temperature increases. We observe an increase in refractive index with the temperature annealing, contrary to the band gap energy which decreases. Annealing has been found to have a strong influence on structural and optical properties of nanocrystalline TiO_2 thin films.

References

[1] Jin P, Miao L, Tanemura S, Xu G, Tazawa M, Yoshimura K 2003 *Appl. Surf. Sci.* **212–213,** 775–781.
[2] Ko H H et al., 2012 *Int. J. Mol. Sci.* **13**, 1658–1669.
[3] Huang L, Liu T, Zheng H, Guo W, Zeng W 2012 *J. Mater. Sci: Mater. Electron* **23,** 2024–2029.
[4] Xu Z, Manhong L, Yongfa Z 2007 *Thin Solid Films* **515,** 7127–7134.
[5] Torchani A, Gharbi R and Fathallah M 2014 *Sensors & Transducers*, **27**, 185-189.
[6] Afuyonia M , Nashed G, Nasser I M 2011 *Energy Procedia* **6**, 11–20
[7] Brinker C J, Harrington M S 1981 *Solar Energy Mater*, **5**, 159 -172.
[8] Bessergenev V G et al. 2002 *Vacuum* **64,** 275–279.
[9] Biju K P, Jain M K 2008 *Thin Solid Films* **516**, 2175-2180.
[10] Ranjitha A, Muthukumarasamy N, Thambidurai M, Balasun Daraprabhu R, Agilan S 2013 *Optik* **124**, 6201– 6204.
[11] Sugapriya S, Sriram R, Lakshmi S 2013 *Optik* **124**, 4971– 4975.
[12] Cullity B D 1978, *Elements of X-ray Diffraction*, Wesley, second ed Addison.
[13] Calderon-Moreno J M et al. 2014 *Ceramics International* **40**, 2209–2220.
[14] Vasanthkumar C V R, Mansingh A 1990, in Seventh IEE International symposium on Application of Ferroelectrics**,** *IEE*, New York **713**.
[15] Moss T S 1961 *Optical Properties of Semiconductors*, Butterworths, Sci. Pub. Ltd, London.
[16] Mathews N R et al. 2009 *Solar Energy* **83**, 1499–1508.
[17] Mechiakh R, Ben Sedrine N, Chtourou R 2011 *Applied Surface Science* **257**, 9103-9109.
[18] Khan A F et al. 2014 *Material Science in Semiconductor Processing* **2**, 1369-8001.

Journal of Physics: Conference Series **596** (2015) 012013 doi:10.1088/1742-6596/596/1/012013

Fresh water production from municipal waste water with RO membrane technology and its application for agriculture and industry in arid area

F Yokoyama[1,2]
[1]Alliance for Research on North Africa (ARENA), University of Tsukuba, Japan
[2]Department of Chemistry, Faculty of Science, Josai University, Japan

E-mail: yokoyamaf@aol.com

Abstract. One of the biggest problems of the 21st century is the global water shortage. Therefore it is difficult to increase the quantity of conventional water resources such as surface water and groundwater for agriculture and industry in arid area. Technical advancement in water treatment membrane technology including RO membrane has been remarkable especially in recent years. As the pore size of RO membrane is less than one nanometer, it is possible to produce the fresh water, which satisfies the drinking water quality standards, with utilizing RO membrane. In this report a new fresh water resource from municipal waste water is studied to apply to the plant factory which is the water saving type agriculture and industry in arid area.

1. Introduction

According to the report [1] of the United Nations in 1997, 97.5% of the water on the earth is sea water, while the remaining 2.5% is fresh water. The source of drinking water in fresh water, such as river, lake, pond and groundwater, constitutes only 0.8% of the entire water resources of the earth. On the contrary the world population will reach 8 billion in 2025 according to WHO [2].

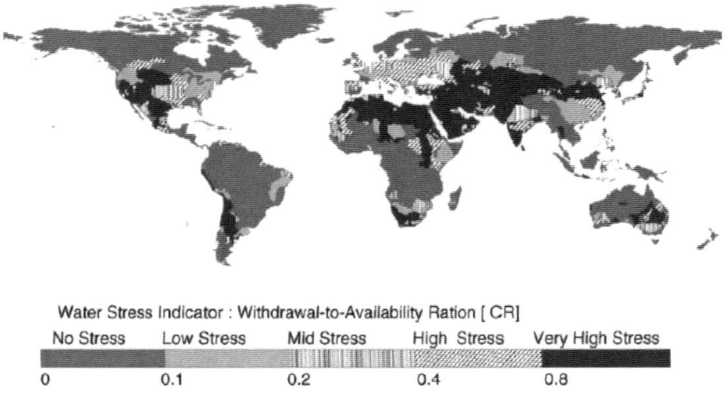

Figure 1. Prediction of global water shortage in 2025 (Water stress map) [3].

At that time half of the world population will live in the countries which suffer from water shortage because of increasing demand of drinking water by increasing population, increasing demand of agricultural water by increasing food production and increasing demand of industrial water by progress of industry. It is difficult to increase the quantity of surface water and groundwater for agriculture and industry in arid area, because the quantity of them is limited in this area and the priority of drinking water is much higher than agricultural and industrial applications. Therefore it is very much attractive, if the fresh water for agriculture and industry in arid area can be produced with RO membrane from municipal waste water which is now discharged as drain.

2. Classification of water treatment membrane

There are four kinds of water treatment membranes. They are classified as RO (Reverse Osmosis) membrane with the smallest pore size, NF (Nanofiltration) membrane, UF (Ultrafiltration) membrane and MF (Microfiltration) membrane. The separation for MF and UF membranes is based on sieve filtration technology, whereas for RO and NF it is based on reverse osmosis technology.

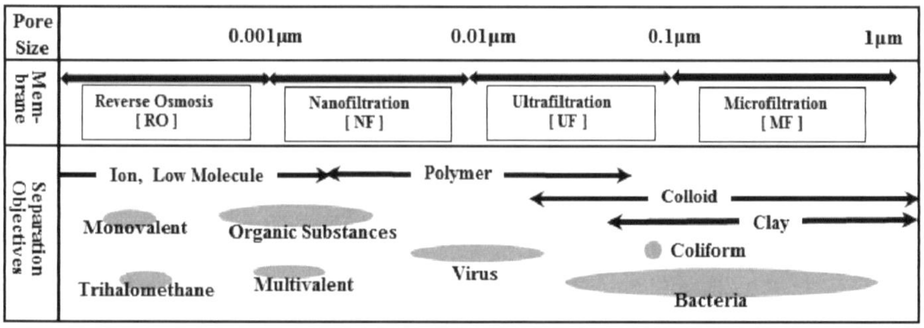

Figure 2. Classification of water treatment membrane.

3. Reverse osmosis membrane and its element structure

3.1. Structure of RO membrane and its element

There are two kinds of RO membrane form which are asymmetric and composite. Historically

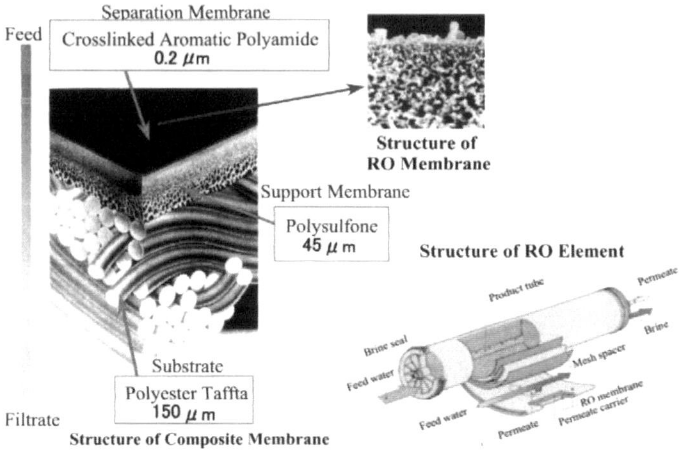

Figure 3. Structure of composite type RO membrane and spiral wound element.

asymmetric membrane of cellulose acetate was first developed and asymmetric membrane of liner polyamide followed. Later, a thin film composite membrane has been developed with different materials which make a separating functional layer and a supporting layer. The majority of RO membranes, which are currently used, is a thin film composite membrane. A spiral wound element is made spirally wound with envelopes of composite membranes, mesh spacers and permeate carriers. As configurations of RO elements, there are plate and frame, tubular and hollow fiber types in addition to spiral wound type.

3.2. History of RO membrane

In the 1950s, the research for sea water desalination has started under the support by Office of Saline Water in the United States. In 1960 Loeb and Sourirajan of University from California had developed the RO membrane of cellulose acetate which was the starting point of water treatment membrane progress.

Fortunately Shoji Kimura of University of Tokyo [4] and Haruhiko Ohya of Yokohama National University [5] had studied and researched jointly on RO membrane under Sourirajan and brought the knowledge back to Japan in the 1960s and the early 1970s. This is one of the reasons why the research on RO membranes in Japan has been kept the highest level in the world. It was very fortunate for Japanese membrane manufacturers to keep their research and technical levels of RO and other membranes globally high from the early stages. There are limited to only two countries of Japan and USA in the world for a long period who can produce spiral wound and hollow fibre types of RO modules, which are mainly applied to water treatment application.

3.3. Technical trend of RO membrane [6]

The trend in the development of RO membrane technology has three directions, shown as Figure 4.

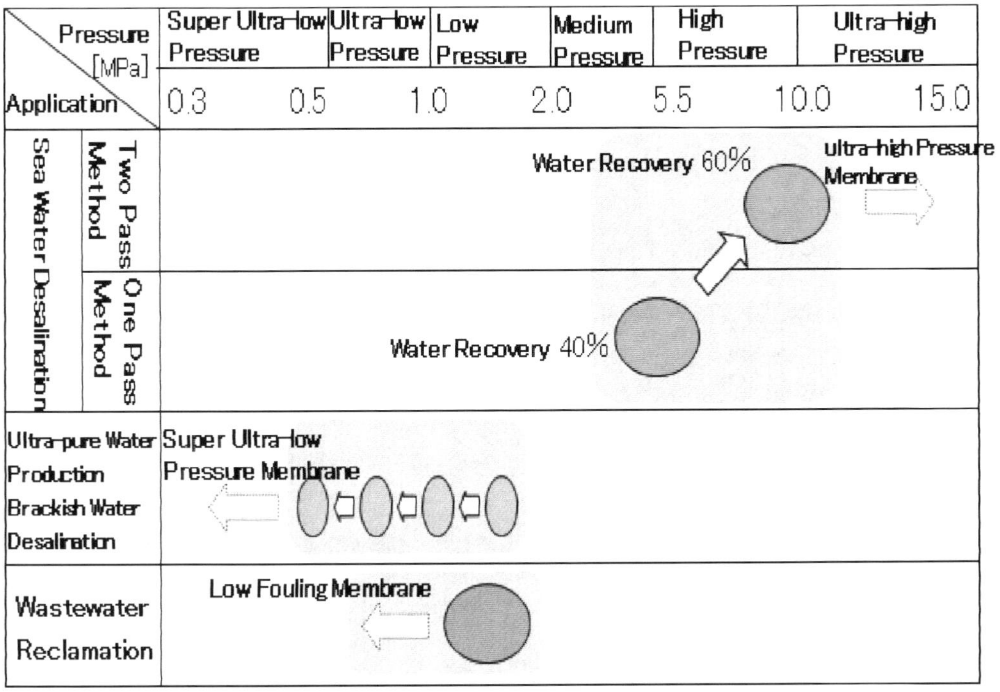

Figure 4. Technical trend of RO membrane.

One is the direction of development for sea water desalination under high salinity. In this area to reduce water production cost it is required to increase water recovery ratio from 40% of the conventional value to 60%. The development of RO membrane and element structure has been done for high pressure durability under more than 10MPa.

Second direction is the development of higher performance under a low salinity area for ultrapure water production and brackish water desalination. In this area to reduce operating cost it is required to increase water productivity under low pressure of less than 1MPa with keeping high salt rejection.

Third direction is the development of durability against membrane fouling for stable RO operation and to expand the application of waste water treatment.

3.4. Application of RO membrane

Reverse osmosis membrane is widely used as an important water treatment technology for production of ultrapure water, boiler feed water and drinking water with desalination of brackish water and sea water. As was mentioned in the beginning one of the answers to improve water shortage in the 21st century is production of drinking water from sea water with RO membrane technology. And it is expected that application of RO membrane will be expanded significantly in the near future.

4. Fresh water production from municipal waste water with membrane technology

4.1. Water resources for agriculture in arid area

Thinking about potential water resources of agricultural and industrial water in arid area, there are three kinds of water such as surface water of river, lake and pond, groundwater and sea water. Surface water and groundwater have been already used for agriculture and industry. Table 1 shows the potential water resources in arid area.

Table 1. Potential water resources in arid area.

Water Resource	Traditional Main Application	Quantity in Aread Area	Treatment Cost	New Water Resource Potentiality
Surface Water •River Water •Lake Water •Pond Water	•Drinking Water •Agriculture	Shortage	Reasonable	Difficult
Groundwater	•Drinking Water	Shortage	Reasonable	Difficult
Sea Water	•Drinking Water	Enough	Expensive	Low
Municipal Waste Water	•Drainrage Discharge	Enough	Reasonable	High

It is difficult to increase the quantity of surface water and groundwater for agriculture and industry in arid area, because the quantity of surface water and groundwater is limited in this area, and the priority of drinking water is much higher than that of agricultural and industrial water in arid area.

Though there is not the quantity problem for agricultural and industrial water from sea water, the cost of sea water desalination is much higher than those of surface water treatment and groundwater desalination, because salinity of sea water is extremely high ranging from 3.5% to 4.5%, comparing to those of surface water and groundwater which are normally less than 1%. Therefore it is very much attractive, if the fresh water for agriculture and industry in arid area can be produced from municipal waste water with water treatment membrane which is discharged as drain. The quantity of municipal waste water is in proportion to population and stable. The cost of municipal waste water treatment is much lower than that of sea water desalination.

4.2. Progress of municipal waste water reclamation process [7]

4.2.1 Advanced municipal waste water reclamation process (1st Step)
In the first step of advanced waste water reclamation process, the complicated pretreatment process with chemical and physical methods was adopted in order to get a high quality RO feed water as shown in Figure 5. The world famous Water Factory 21 at Orange County Water District is the real pioneer of advanced municipal waste water reclamation process in the world. Water Factory 21 has started its operation in 1979. The treated water from activated sludge tank was fed to RO process after lime clarification, re-carbonation and sand filtration. RO permeate was injected into the ground after chlorine disinfection. The water quality after treatment in Water Factory 21 met and exceeded the drinking water quality standards in USA.

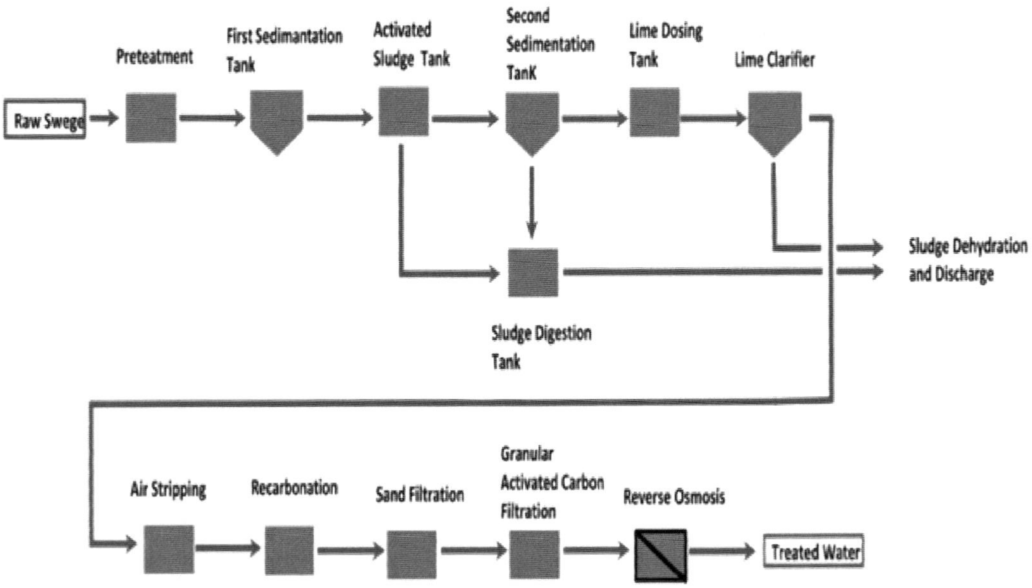

Figure 5. Advanced municipal waste water reclamation process (1st step).

4.2.2. Advanced municipal waste water reclamation process (2nd step) - Introduction of MF/UF membrane
Thanks to introduction of MF (UF) membrane, the pretreatment system for removal of suspended materials had become simple. Water after treatment in sedimentation tank from activated sludge tank was fed to RO system after treating with MF (UF) membrane system and removing suspended materials. RO permeate can be reused. The NEWater plant by PUB (Public Utility Board) in Singapore is one of the most recognizable examples of the 2nd step of advanced municipal waste water reclamation process. The NEWater plant was developed by modelling after Orange County

Water District Water Factory 21. The feature of this process was the adoption of MF membrane process instead of the complicated pretreatment process with chemical and physical methods, thus the total process became extremely simple and economical. PUB named their treated water "NEWater" for distinguishing it from other treated water from waste water.

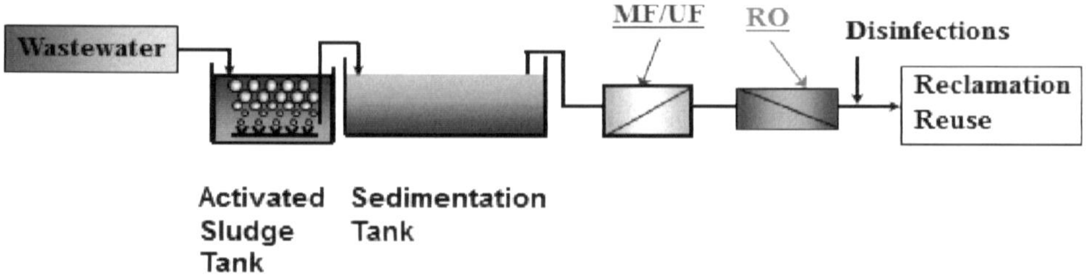

Figure 6. Advanced municipal waste water reclamation process with MF/UF before RO (2nd Step).

The water quality of NEWater can easily clear the WHO drinking water quality standards. Singapore government distributed more than 1,000,000 bottles of NEWater at school sports festivals and on National Holidays in Singapore in order to capture people's attention.

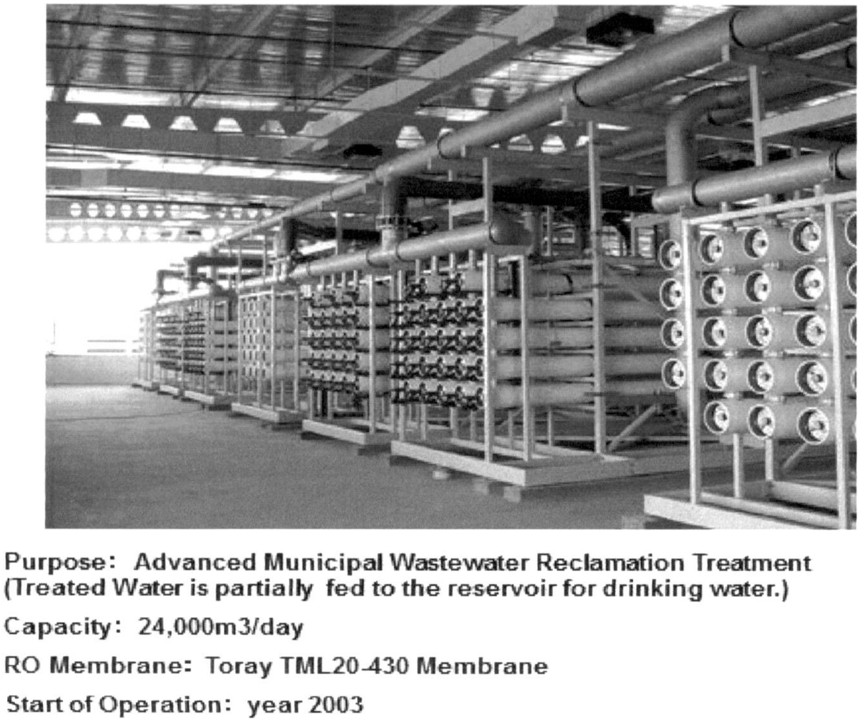

Figure 7. Singapore Seletar NEWater RO plant.

4.2.3. Advanced municipal waste water reclamation process with MBR before RO (3rd step)

With adoption of the Membrane Bio-Reactor (MBR) that combines activated sludge tank and MF process, the total process became simpler and more economical. In 2009 the Changi NEWater plant of the advanced municipal waste water reclamation process with MBR began its operation.

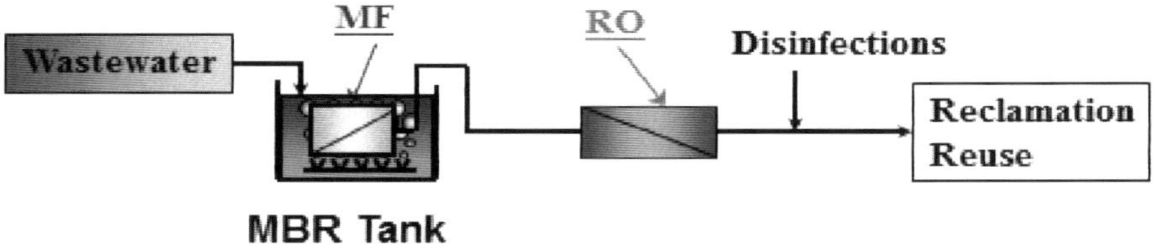

Figure 8. Advanced municipal waste water reclamation process with MBR before RO (3rd Step).

5. Agricultural application of fresh water from municipal waste water

5.1 Plant factory – water saving type agriculture

It is mentioned that in arid area fresh water from municipal waste water with membrane technology is expected as a new water resource for agriculture. The figure 9 shows the water consumption of agriculture by cultivation method [8],[9].

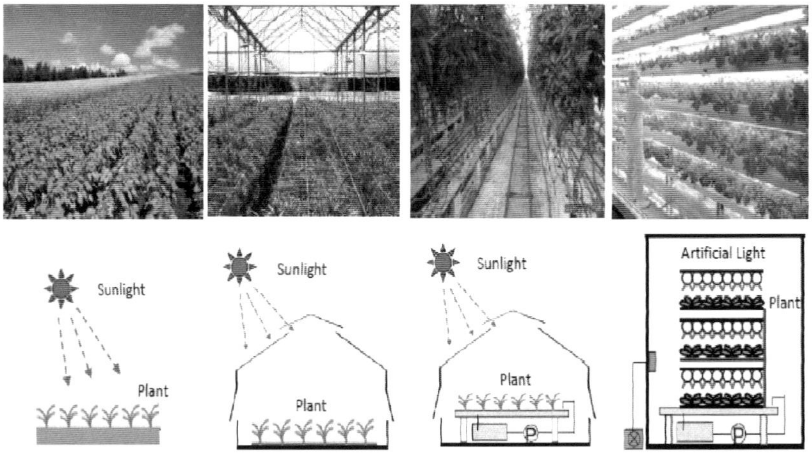

Cultivation System	Dry-field	Green House (Soil Culture)	Sun Light Type Plant Factory (Hydroponics, Open)	Artificial Light Type Plant factory (Hydroponics, Closed)
Water Vaporization from Soil	O	O	x	x
Water Penetration into Soil	O	O	x	x
Water Vaporization from Leaves	O	O	O	x (Recovery of Vaporized Water)
Water Consumption Ratio	1	1	1/2	1/20

Figure 9. Water consumption by cultivation method.

If the water consumption ratio of dry-field farming is 1, then the water consumption ratio of green house with soil cultivation is 1 – 1/2. The water consumption ratio of sunlight type plant factory with hydroponics is 1/2 – 1/10, because there is no water vaporization from soil and no water penetration into soil. The water consumption ratio of artificial light type plant factory decreases 1/10 – 1/100, because vaporized water from leaves is recovered by dehumidification with air conditioner.

5.2 Integration of new water resource, agriculture and photovoltaic generator [10],[11],[12]
In arid area it is difficult to increase quantity of surface water and groundwater for agriculture, because their quantity is limited. It is also difficult to apply the fresh water from sea water to agriculture because its desalination cost is high. Therefore it is very much attractive as a new water resource of agriculture in an arid area, if the fresh water of drinking water quality level can be economically produced from municipal waste water, which is discharged as drain, with advanced RO membrane technology. The new agricultural method in arid area is proposed as follows:

- Photovoltaic (PV) generator is applied for energy resource.
- Fresh water from municipal waste water with RO membrane technology is applied for agricultural water.
- Plant factory of water saving agriculture is applied for cultivation method.

Figure 10 shows a conceptual figure of "Fresh water production from municipal waste water with RO membrane technology and its application for plant factories with Photovoltaic (PV) generator".

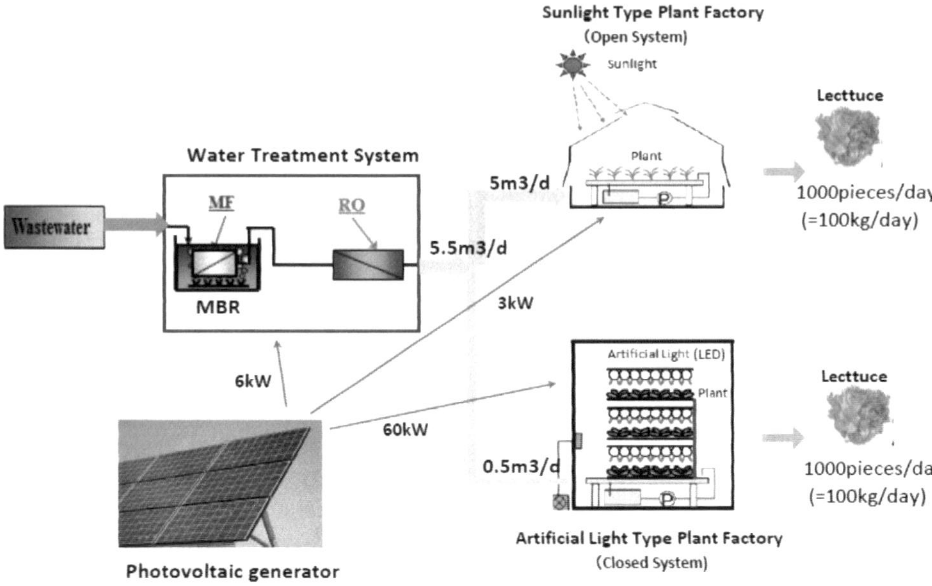

Figure 10. Fresh water production from municipal waste water with RO membrane technology and its application for plant factories with PV generator.

Journal of Physics: Conference Series **596** (2015) 012013 doi:10.1088/1742-6596/596/1/012013

6. Industrial application of fresh water from municipal waste water

In arid area it is difficult to increase quantity of industrial water and in order to promote industries it is necessary to snatch existing water resources for drinking and agriculture against people and farmers. If the fresh water, whose quality is equal to or better than WHO drinking water quality standards, can be produced from municipal waste water with RO membrane technology, it is not necessary to snatch existing water resources. The fresh water from municipal waste water can be used for industrial water, promote industries and create many new jobs.

7. Conclusion

Municipal waste water is abundant and stable in proportion to population. Thanks to advanced RO (Reverse Osmosis) membrane technology it is possible to produce the fresh water, which satisfies WHO drinking water quality standards, from municipal waste water. Therefore municipal waste water becomes a potential water resource for agriculture and industry in arid area. The plant factory of hydroponic cultivation system with photovoltaic generation is a water saving type agriculture and suitable in arid area.

References

[1] Korzoun V I and Sokolov A A 1997 *World Water Balance and Water Resources of the earth E/Conf, 70 /TP127* (Presentation Document at the United Nations Water Conference)
[2] World Health Organization 2008 *10 facts about water scarcity*
[3] Rijsberman R F 2006 Water scarcity: Fact or fiction? *Agr. Water Manag.* **80** 5-22
[4] Kimura S and Nakao S 1997 *Separation Technology-Membrane Separation* (Tokyo: Dainihon Toshyo)
[5] Ohya H 1987 *Membrane "Application Technology Handbook II Reverse Osmosis Membrane Method Ultrafiltration Method* (Tokyo: Saiwai-Shyobou)
[6] Fusaoka Y and Kurihara M 2002 *Proceedings of IDA World Congress* (Bahrain).
[7] Yokoyama F 2006 *MRC News* vol 35 (Tokyo) p41-48 .
[8] Kozai T 2012 *Plant Factory with Artificial Light* (Tokyo: Ohmsha) p28
[9] Goto E 2012 *2nd Int. Symp. of Dev. Trend* (Taipei)
[10] Yokoyama F 2013 *Proc. of 3rd Asia-Arab Sustain. Energ. Forum*, (Hirosaki: Japan)
[11] Yokoyama,F 2013 *Proc. of the Japan-Tunisia Symp. on Science, Society & Technology* (Hammamet: Tunisia).
[12] Stambouli A, Yokoyama F, et al, 2014 *Renew. and Sustain. Energ. Rev.* **30** 912

Electrochemical detection of phenolic estrogenic compounds at clay modified carbon paste electrode

N Belkamssa[1], L Ouattara[2], A Kawachi[3], M Tsujimura[3], H Isoda[3], A Chtaini[4] and M Ksibi[1]

[1]Laboratory Water Energy and Environment L3E-ENIS, University of Sfax, Route Soukra Km 4,BP 1173,3028 Sfax Tunisia.
[2] Laboratory of Physical Chemistry, University UFR SSMT-Félix Houphouët-Boigny-Cocody, 22 BP 582 Abidjan 22 Côte d'Ivoire.
[3]Graduate School of Life and Environmental Sciences, University of Tsukuba, 1-1-1Tennodai, Tsukuba, Ibaraki305-8572, Japan.
[4]Team Molecular Electrochemistry and Inorganic Materials, University Sultan Moulay Sliman, Faculty of Science and Technology, Beni Mellel Morocco.

E-mail : *mohamed.ksibi@tunet.tn*

Abstract. A simple and sensitive electroanalytical method was developed to determine the Endocrine Disrupting chemical 4-tert-octylphenol on clay modified carbon paste electrode (Clay/CPE). The electrochemical response of the proposed electrode was studied by means of cyclic and square wave voltammetry. It has found that the oxidation of 4-tert-octylphenol on the clay/CPE displayed a well-defined oxidation peak. Under these optimal conditions, a linear relation between concentrations of 4-tert-octylphenol current response was obtained over range of 7.26×10^{-6} to 3.87×10^{-7} with a detection and quantification limit of $9.2\ \times10^{-7}$ M and 3.06×10^{-6} M, respectively. The correlation coefficient is 0.9963. The modified electrode showed suitable sensitivity, high stability and an accurate detection of 4-tert-octylphenol.The modified electrode also relevant suitable selectivity for various phenolic estrogenic compounds.

1. Introduction

Since the middle of the last decade, a variety of adverse effects of endocrine disrupting chemicals (EDCs) on the endocrine systems of human and animals have been observed. Alkylphenols are released into the aquatic environment primarily as major metabolites of alkyl phenol ethoxylates. For example, 4-tert-octylphenol is one of the most important substances and intermediates. They are ubiquitous in the environment because of their large domestic, industrial and agricultural use **[1].** Currently LC-MS and HPLC has demonstrated to be a useful technique for the analysis of EDCs in environmental and biological samples **[2]**.In recent years, some phenols have been successfully detected by using electrochemical methods **[3, 4].** However, carbon nanotube (CNT) film was used for the detection of 4-tert-Octylphenol **[5]**. As shown previously, the clay modified electrodes are types of chemical modified electrodes and their potential use for electroanalysis **[6]**. In this study, cyclic and square wave voltammetry on disposable carbon paste electrodes (CPE), bare or modified with clay, were used for direct electrochemical detection of the 4-tert-octylphenol.

---

[1] *mohamed.ksibi@tunet.tn*

2. EXPERIMENTAL

2.1. Reagents
All chemicals were of analytical grade and have been used as received without any further purification. All solutions were prepared in double distilled water. 4-*tert*-Octylphenol was purchased from Sigma-Aldrich (USA) with the purity of 97%. The commercial graphite powder was supplied from France (Carbone, Lorraine, ref 9900). The clay samples employed in the research were taken from the Natural Resources of Cameroon, more precisely from the town of Garoua, in the region of North Cameroon. The samples were washed with deionized water several times with constant stirring. After being dried at 900 °C for 1 hour, the samples were crushed, and finally sieved into different particle size ranging $\leq$ 125 μm. The natural phosphate used in this work is collected from the Province of Khouribga in Morocco.

2.1. Instrument
Cyclic and square wave voltammetry were carried out with a voltalab potentiostat (model PGSTAT 100, Ecochemie B.V., Utrecht, The Netherlands) driven by the general purpose of the electrochemical systems data processing software (voltalab master 4 software). The electrochemical cell was configured to work with three electrodes by using Clay/CPE as the working electrode, platinum plate for counter and saturated calomel electrode(SCE) as reference electrodes. The surface area of the working electrode is about 0.12cm^2.

2.3. Synthesis of working electrode
The carbon paste electrodes were prepared by mixing high purity graphite powder and a Cameroon's clay with paraffin oil. That mixture named clay/CPE was then compacted carefully in the electrode cavity. The electrical connection was provided by a bar of carbon.

2.4. Analytical procedure
The initial procedure consisted in measuring the electrochemical response at the clay/CPE electrode at a fixed concentration of 4-tert-octylphenol. Standard solution of 4-tert-octylphenol was added into the electrochemical cell containing 0.1 M of the supporting electrolyte (Na_2SO_4). The 4-tert-octylphenolcontaining supporting electrolyte was kept 20 s at open circuit before performing the voltammetric measurements. The square wave voltammetry was recorded in the range from -1.5 V to 1.5 V with a scan rate of 50 mV.s^{-1}, a step potential of 50 mV, the amplitude of 2 mV and 0.1s duration. Optimum conditions were established by measuring the peak currents versus various investigated parameters such as preconcentration time, scan rate and concentration. All the experiments were performed at room temperature (25°C).

3. Results and discussion

3.1. Electrochemical impedance spectroscopy (EIS)
Electrochemical impedance spectroscopy (EIS) has been proven to be one of the most powerful tools for probing the features of surface-modified electrodes. EIS measurements, shown in figure 1, were performed in the presence of 26.6 × 10^{-6} M of 4-tert-octylphenol in 0.1 M Na_2SO_4 and the frequency range is from 0.1 to 10^5 Hz. Here Zr and Z are the real and imaginary parts of experimental data. At the higher frequencies, squeezed semicircles represent an electron transfer-limited process, followed by a diffusion ally limited electron transfer process at the lower frequencies **[7]**. The diameter of this semicircle in the Nyquist plot, which exhibits the electron transfer resistance (R_{ct}) of the layer, can be used to describe the interface properties of the electrode for each immobilization step **[8, 9]**. The charge-transfer resistance (R_{ct}) values are calculated from the difference in impedance at lower and higher frequencies, as suggested by Tsuru et al **[10]**. To obtain the double layer capacitance (C_{dl}) the

frequency at which the imaginary component of the impedance is maximal (-Zmax) is found as represented in the equation (1):

$$C_{dl} = \frac{1}{\omega R_{ct}} w\ here \omega = 2\pi f_{(-Zmax)} \quad (1)$$

With $f_{(-Zmax)} = 100$ mHz

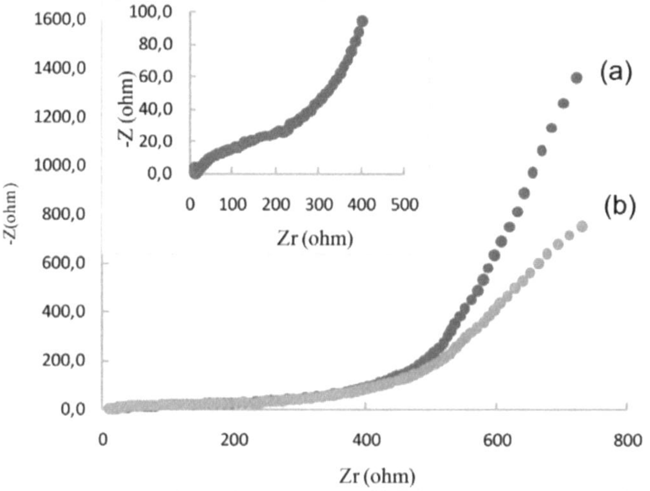

Figure 1.Nyquist plots obtained for Clay/CPE (a), Clay/CPE/4-tert-octylphenol (b). EIS measurements were carried out in 26.6 x 10^{-6}M of 4-tert-octylphenol containing 0.1 M of Na_2SO_4

The impedance diagrams were obtained for the frequencies ranging from100 kHz to 100 mHz. The values of the charge-transfer resistance (R_{ct}) and those of double-layer capacitance (C_{dl}) are given in table 1. As shown in Figure 1, A semicircle with a large diameter is observed at the unmodified electrode, indicating the high charge resistance of CPE (189 Ω.cm²) and it is difficult for the electron transfer at electrode surface. However, the diameter of semicircle diminishes significantly at clay/CPE/4-tert-octylphenol. It was obvious that charge transfer resistance (140 Ω.cm²) of the electrode surface decreased, which indicated that the presence of products exhibited good conductivity and the electron transfer rate at the electrode/solution interface was greatly increased. This result is similar to that obtained by Xia Li et al **[11]**. Further, the C_{dl} value for the Clay/CPE/4-tert-octylphenol electrode is much higher than that for the CPE revealing that the 4-tert-octylphenol can easily diffuse through the clay/CPE electrode layer.

Table 1. Electrical parameters calculated from the impedance spectra carried out in 0.1 M Na_2SO_4on the clay/CPE and clay/CPE/4-tert- octylphenol.

Electrode	R_{ct} (Ω.cm²)	C_{dl}(μF/cm²)
Clay/CPE	189	42.4
Clay/CPE/4-tert-octylpehnol	140	35.1

3.2. Cyclic voltammetry measurements

The cyclic voltammograms have been recorded Clay/CPE. Measurements have been performed with the modified carbon paste electrodes before and after incubation with 4-tert-octylphenol. The results are presented in Figure 2. In the blank electrolyte i.e. in absence of 4-tert-octylphenol precisely in the electrolyte, no anodic oxidation peak is observed on clay/CPE electrodes. Only an anodic peak on clay/CPE at ~0.75V is seen after adding the 4-tert-octylphenol.

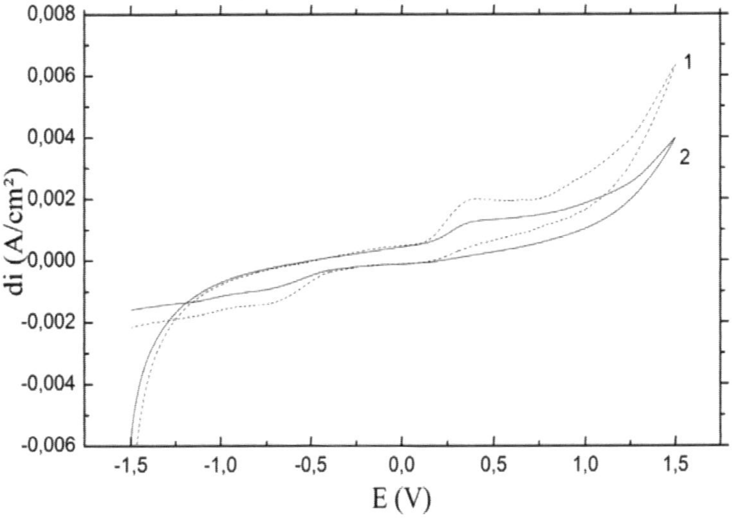

Figure 2.Cyclic voltammograms recorded on the clay/CPE containing 0.1 M Na_2SO_4 at a scan rate of 50 mV/s: (1) in absence and (2) in presence of 4-tert-octylphenol.

3.3. Calibration curve

The square wave voltammetry was performed in figure 3 and a well-defined 4-tert- octylphenol's oxidation peak was obtained. In its inset, a linear evolution of the peak current density towards the 4-tert-octylphenol concentration was reached. As it is obtained, the oxidation peak current (Ip) is proportional to 4-tert-octylphenol concentration in the range of 7.26×10^{-6} M to 38.7×10^{-6} M with a detection and quantification limit of 9.2×10^{-7} M and 3.06×10^{-6} M, respectively.

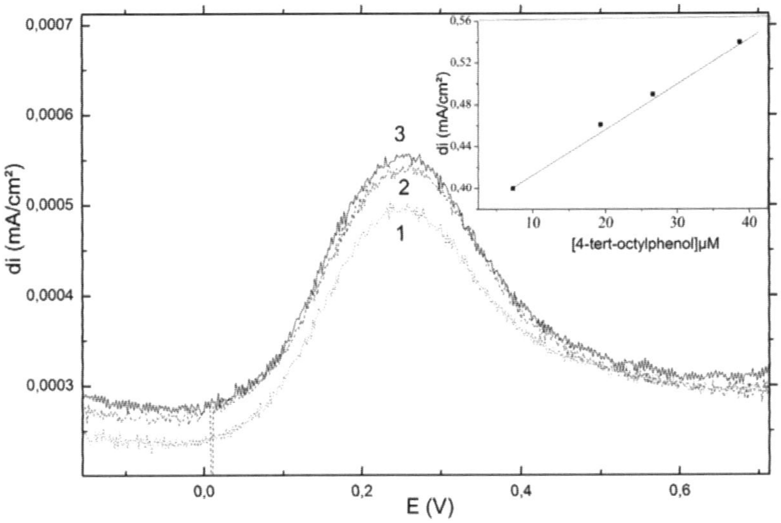

Figure 3. Square wave voltammograms at clay/CPE in 0.1 M Na_2SO_4 for different concentrations of 4-tert-octylphenol; (1) 7.26×10^{-6} M, (2) 19.3×10^{-6} M, (3) 26.6×10^{-6} M.

Journal of Physics: Conference Series **596** (2015) 012014 doi:10.1088/1742-6596/596/1/012014

The relative standard deviation is less than 3.8 % for 26.6 × 10 $^{-6}$ M 4-tert-octylphenol (for four analysis), which shows good reproducibility.

3.4. Comparison of the activities of the prepared electrode towards several endocrine disrupting chemicals

The comparison of a cyclic voltammograms obtained at a Clay/CPE; a natural phosphate /CPE at an unmodified CPE for 26.6×10^{-6} M solutions of 4-tert-Octylphenol, 4-Nonylphenol and Bisphenol A shown, respectively, a well-defined oxidation peaks at 0.5V, 0.35 V and at 0.33V at the Clay/CPE, whose peak currents are considerably larger than those obtained at the bare CPE with the exception of 4-tert-octylphenol. These results indicated the electrocatalytic activity of the clay modified carbon paste electrodes towards all the estrogenic compounds. The relative standard deviation (RSD %) (S/N=3) is 2.04 % for 1.81× 10 $^{-6}$ M 4-nonylphenol and 3.4 % for 8.76× 10 $^{-6}$ M Bisphenol A. So the repeatability of electrode like its reproducibility is good.

4. Conclusion

In the current study, a simple and sensitive electrochemical method was proposed for the determination of 4-tert-octylphenol based on Clay/CPE. The modification of the carbon paste electrode with clay produced, an enhancement of the electrochemical oxidation responses obtained for phenolic estrogenic compounds.

References

[1] Schilirò T, Pignata C, Rovere R, Fea E, Gilli G 2009 *Chemosphere.* **75** 335-340
[2] Bellfroid A C, Van der Horst A, Vethaak A D, Schafer A J, Rijs G B J, Wegener J , Cofino W P 1999 *Sci. Total. Environ.* **225** 101
[3] Yang C 2004 *Microchim. Acta.* **148** 87
[4] Huang W, Yang C, Zhang S 2003 *Anal. Bioanal. Chem.* **375** 703
[5] Qijin W, Ping Y, Hui C, Hongtu S, Nianjun Y 2013 *Electroanalyt. Chem.* **689** 252-256
[6] Stalder C J, Chao S, Summers D P, Wrighton M S 1983 *J. Am. Chem. Soc.* **105** 6318
[7] Billah M, Hays H C W, Millner P A 2008 *Microchim. Acta.* **160** 447
[8] Lin C C, Chen LC, Huang C H, Ding S J, Chang C C, Chang H C 2008 *J. Electroanal. Chem.* **39** 619-620
[9] Darain F, Park D S, Park J S, Shim Y B 2004 *Biosens. Bioelectron.* **19** 1245
[10] Tsuru T, Haruyama S , Gijutsu B 1978 *J. Jpn .Soc. Corros. Eng.* **573** 27
[11] Li X, Niu Q, Li G, Zhan T, Sun W 2011 *J. Braz. Chem. Soc.* **22** 422

Direct Carbothermic Silica Reduction from Purified Silica to Solar-Grade Silicon

Kenji Itaka[1]*, Takuya Ogasawara[1], Abderahmane Boucetta[1], Rabie Benioub[1], Masatomo Sumiya[2], Takuya Hashimoto[3], Hideomi Koinuma[4], Yasubumi Furuya[1]

[1] North Japan Research Institute for Sustainable Energy, Hirosaki University, 2-1-3, Matsubara, Aomori, 030-0813, Japan.
[2] Widegap Materials group, National Institute for Materials Science, Tsukuba, 305-0044,Japan.
[3] College of Humanities and Sciences, Nihon University, 3-25-40 Sakurajousi, Setagaya-ku, Tokyo 156-8550, Japan.
[4] Graduate School of Frontier Sciences, The University of Tokyo, Kashiwa, 277-8568,Japan.

*E-mail: itaka@cc.hirosaki-u.ac.jp

Abstract. The shortage of silicon for solar cells has much attract in research and development of the low cost mass production process of solar grade Silicon (SOG-Si). The key of the strategy to silicon for solar cells in long terms is the efficiency in reduction. We developed a new concept reduction furnace using the combinatorial method in order to investigate the metallurgical process for silicon reduction. The production of reduced Si shows purity with less 2ppm level.

1. Introduction

Two peculiar features of deserts, which are powerful solar radiation and non-use vast area, represent the treasury of solar energy that is one of candidates for the next generation sustainable energy from Nature[1, 2]. In viewpoint of reserves of natural resources, the material selection for solar cells which covers the vast deserts is only the silicon, which exists generally in large quantities as oxides in the desert as element on the earth. The Siemens method, which has been a silicon reduction process used for many years, is suitable for the integrated semiconductor industry which needs ultra-high purity grade. However, the cost of this process is too high due to low yield in the reaction process of chloride gas. The purity level of silicon for the solar cells is required to be 6N (99.9999%) with respect to the impurities of boron

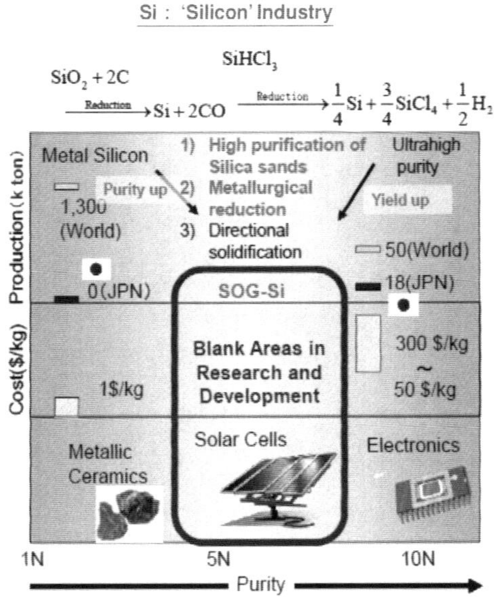

Figure 1. Strategy to 'Silicon' business industry.

Journal of Physics: Conference Series **596** (2015) 012015 doi:10.1088/1742-6596/596/1/012015

and phosphorus. This purity grade is much higher than that of metal silicon (~2N), and much lower than that of semiconductor silicon (~11N) as shown in Figure 1. The shortage of silicon for solar cells has much attract in research and development of the low cost mass production process of SOG-Si. There are two strategic paths to the process of 6N-graded silicon, that is, solar-grade silicon (SOG-Si): One is the way to improve the process of metal silicon in purity using high-purity silicon dioxides. The other is the way to improve the yield of the process of the silicon for electronics.

Figure 2 shows schematic process charts of our direct oxygen-reduction process compared with conventional Siemens process. In our innovative and challenging approach, the one of key processes is the reduction process from high-purity silicon dioxide to high purity silicon using high-purity granular SiO_2 powder and carbon[3]. This process has three following merits: (1) we can use not silica stones, but silica sands. (2) low-energy consumption process using chemical process for purification of silica.(3) Halogen-free process.

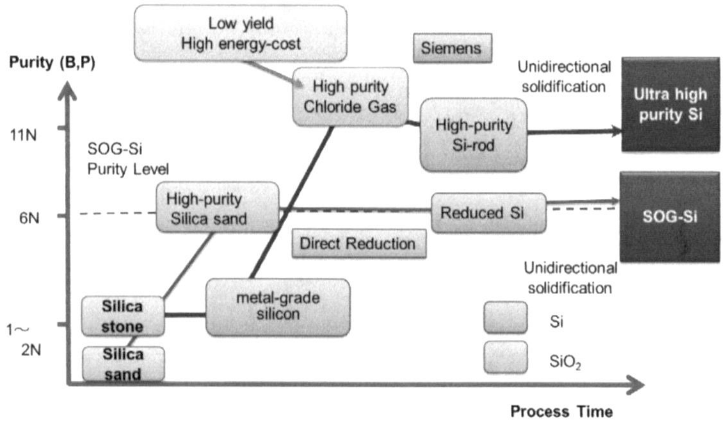

Figure 2. Process charts of direct oxygen-reduction process and Siemens process.

2. Thermodynamics in Direct Reduction of Silica

Reduction and synthesis processes of oxides can be roughly estimated with the Ellingham Diagram, which is temperature dependence of chemical potential of oxygen molecules. Figure 3 shows the Ellingham Diagrams of Silicon, Iron, Titanium and Aluminium, which are calculated using thermodynamic database MALT2[4]. Red lines indicate the chemical potential of oxygen molecules in hydrogen reduction and carbon-carbon monoxide reduction processes. The line of the hydrogen reduction crosses only the Iron curve, while the line of the carbon hydrogen reduction crosses the silicon curve. This indicates that the reduction of silicon from silica cannot be performed by normal hydrogen gas. The reduction of silicon has similarities to that of titanium oxides and aluminium oxides.

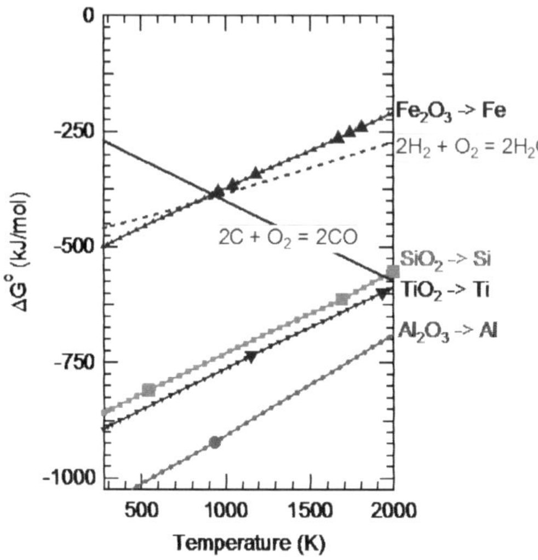

Figure 3. Ellingham Diagrams of Silicon, Iron, Titanium and Aluminum.

In the case of use of carbon for the reduction, however, this reduction reaction consists of a lot of complicated reaction paths as shown in Figure 4. For the optimization of not only the main process but also sub process, the control of many parameters such as reaction temperature, partial pressure and composition is required.

First, SiO_2 reacted to carbon above 1300°C generates SiO gas, which is volatile under conventional condition. Then, the following reactions will proceed.

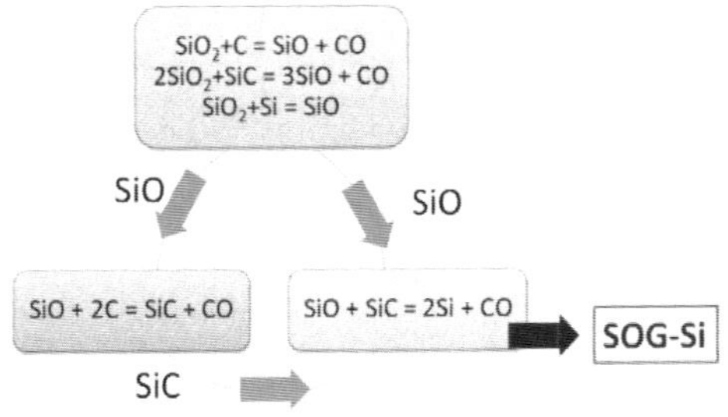

Figure 4. Reaction process paths from SiO_2 to Si.

$$\mathrm{Si}[sl]+\mathrm{SiO_2}[sl]=2\mathrm{SiO}[g] \quad (1)$$

$$\mathrm{SiO}[g]+\mathrm{SiC}[s]=2\mathrm{Si}[sl]+\mathrm{CO}[g] \quad (2)$$

$$\mathrm{SiO}[g]+\mathrm{C}[s]=\mathrm{Si}[sl]+\mathrm{CO}[g] \quad (3)$$

$$\mathrm{SiO}[g]+2\mathrm{C}[s]=\mathrm{SiC}[s]+\mathrm{CO}[g] \quad (4)$$

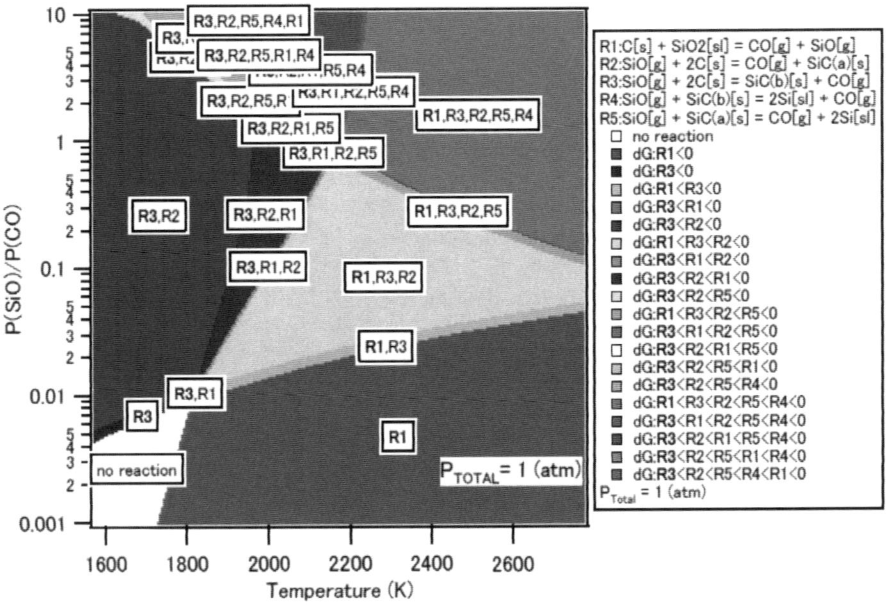

Figure 5. (online color) Phase diagrams of the silica reduction processes. Each bold annotation indicates the preferential reaction in the phase.

Figure 5 shows phase diagrams of the silica reduction processes (a) with carbon and (b) without carbon (only with silicon carbides) calculated using thermodynamic database MALT2[4]. The reaction with the formation of the final product silicon is only R2. What should be noted is that there is no phase with the preferential reaction (2) in Fig. 5 (a). After exhausting carbon, the phase diagram Fig. 5(a) migrates to the phase diagram Fig.5 (b). On the phase diagram Fig. 5(b), the R2 reaction becomes dominant in the region of high SiO partial pressure based on carbon monoxides partial pressure.

3. Combinatorial Silica Reduction Furnace

This reduction reaction consists of a lot of complicated reaction paths as shown in Fig. 4. For the optimization of not only the main process but also sub process, the control of many parameters such as reaction temperature, partial pressure and composition is required. Combinatorial technology is widely applicable to a variety of materials and processes. Ever since the proposal of combinatorial solid state nano science and technology, several automated parallel synthesis systems for high-throughput screening of new functional materials have been developed[5], and the combinatorial autoclave reactor was developed to investigate optimum reaction conditions efficiently for alternating CO_2 copolymerization with epoxide[6]. Thus, the combinatorial methods are powerful tools to optimize a complicated chemical reaction. We developed a new concept reduction furnace using the combinatorial method in order to investigate the metallurgical process for silicon reduction as shown in Figure 6.

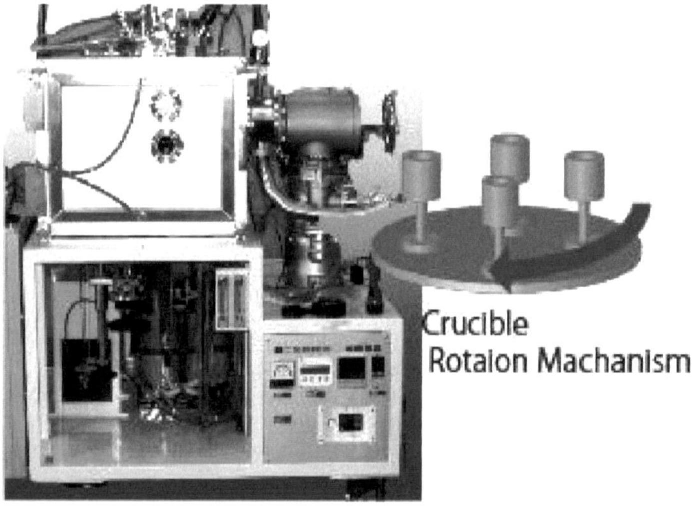

Figure 6. Photo image of combinatorial reduction furnace system.

Our system can run four processes with starting compositions and process temperatures by only one setup. Normally, vacuum evacuation of the chamber takes 2-3 hours using the diffusion pump. In this system, the waste time for vacuum evacuation for each crucible can be saved. In addition, the systematic data can be obtained under similar background condition because of less perturbation. The combination method is useful in systematic screening in these parameters in the optimization. Our developed silica reduction furnace system will serve as a powerful tool for the investigation and optimization in the silica reduction process.

Figure 7 shows the image of the obtained Silicon (weight ~5g) using our furnace. The heating time was 30 min. The impurities level of this sample were 1.1~1.4 ppm for boron and 1.1~2.6 ppm for phosphor.

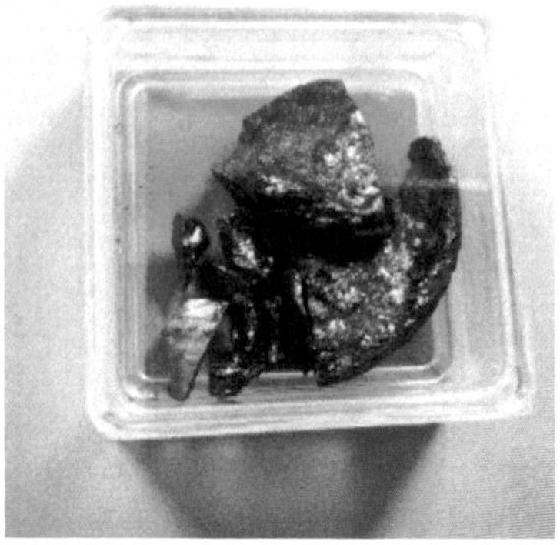

Figure 7. Photo of the obtained Silicon. The impurities level of this sample were 1.1~1.4 ppm for boron and 1.1~2.6 ppm for phosphor.

4. Conclusion

We developed a new concept reduction furnace using the combinatorial method in order to investigate the metallurgical process for silicon reduction. Our system can run four processes with starting compositions and process temperatures by only one setup. The combination method is useful in systematic screening in these parameters in the optimization. Our developed silica reduction system will serve as a powerful tool for the investigation and optimization

Journal of Physics: Conference Series **596** (2015) 012015 doi:10.1088/1742-6596/596/1/012015
in the silica reduction process. We obtained reduced Si with a purity with less 2 ppm level by new direct carbothermic reduction process. The addition of SiC raw materials accelerates the production of silicon.

Acknowledgments

This work was partially supported by Grants-in-Aid for Scientific Research, KAKENHI 23656598. This work was supported by JST-JICA SATREPS program. This work was partially supported by JSPS KAKENHI Grant Numbers 23656598, 23560001. The author would like to thank Prof. H. Sato who are working collaboratively.

5. REFERENCES

[1] Komoto K, Ito M, Van Der Vleuten P, Faiman D and Kurokawa K eds 2009 *Energy from the Desert, Very Large Scale Photovoltaic Systems: Socio-economic, Financial, Technical and Environmental Aspects*: Earthscan Pubns Ltd; Volume Three)

[2] Kurokawa K ed 2006 *Energy from the Desert: Practical Proposals for Very Large Scale Photovoltaic Systems*: Routledge)

[3] Sakaguchi Y, Ishizaki M, Kawahara T, Fukai M, Yoshiyagawa M and Aratani F 1992 Production of High-Purity Silicon by Carbothermic Reduction of Silica Using Ac-Arc Furnace with Heated Shaft *Isij International* **32** 643-9

[4] Yokokawa H, Yamauchi S and Matsumoto T 1994 Thermodynamic Database Malt2 and Its Applications to High-Temperature Materials Chemistry *Thermochimica Acta* **245** 45-55

[5] Koinuma H, Itaka K, Matsumoto Y, Yoshida Y, Aikawa S and Takeuchi K 2010 Vacuum and Pressured Combinatorial Processings for Exploration of Environmental Catalysts *Topics in Catalysis* **53** 35-9

[6] Koinuma H and Takeuchi I 2004 Combinatorial solid-state chemistry of inorganic materials *Nature Materials* **3** 429-38

Performance study of a heat recovery desiccant cooling system

Seifennasr Sabek[1] , Kaouther BenNasr[2], Ridha Chouikh[3] and Amenallah Guizani[1]
[1] Centre des Recherches et des Technologies de l'Energie (CRTEn) PB 95, Hammam Lif 2050, Tunisie.
[2] Ecole Nationale d'Architecture et d'Urbanisme, Carthago 2026, Tunisie.
[3] Institut Supérieur de Mathématiques Appliquées et d'Informatique (ISMAI), Kairouan 3100, Tunisie.

E-mail: Seifennasr.SABEK@gmail.com

Abstract: The comparison between the experimental and theoretical simulations of a desiccant cooling system under various climatic conditions (outdoor temperature and relative humidity) on the system performance has been presented. The performance of the system is evaluated using Cooling Capacity (CC) parameter. The system under a typical summer day of hot and humid climate was tested. A remarkable decrease about 40-65% in the specific humidity and with a supply air temperature lower than 25°C of the proposed system was observed. The study is important and helpful to improve the effectiveness of this kind of liquid desiccant system in hot and humid places.

1. Introduction

The development of air conditioning to improve the living conditions leads to an increase in energy consumption with problematic effects on the environment. Therefore, in recent years the liquid desiccant cooling technology has been studied, focusing on applications such as cooling of human houses, commercial buildings and hospitals.

Liquid desiccant cooling system (LDC) is operated based on liquid desiccant, which has strong affinity to water to control air temperature, reduce humidity and provide high quality air. There are a lot of studies on the performance of LDC systems in different climatic conditions are carried.

Some of these studies have been carried out with hybrid liquid desiccant systems which eliminate the latent heat of vapor condensation before entering in the conventional air conditioning systems. Ma et al. [1] studied a combining system the system of desiccant cooling liquid and a solar system of adsorption.

Hellmann and Grossman investigated a novel cooling system which used a renewable energy such as solar energy [2].

Mei and Dai [3] gave a detailed account of the general features of the major desiccant dehumidification techniques and configurations of the related systems. They concluded that this kind of systems can realize the independent control of process air humidity and temperature, but it has the potential drawbacks which are big dimension and unstable operation.

Other systems comprise a dehumidifier and a regenerator working together to dehumidify and cool the fresh air entering to buildings. The liquid desiccant solutions are used a fluid of operation. Shuli Liu [4] studied a novel heat recovery/cooling system leads by a solar collector and a turn of cooling to carry out low energy cooling with the low emission of CO_2. This system is very useful for a hot and wet climate with the abundance of solar energy.

Grossman and Gommed [5,6] studied and examined a desiccant cooling system for solar cooling and dehumidification, the system including a novel Heat/Mass Exchanger (HME), is designed to be used as desiccant reserve of solution with the regenerator and the dehumidifier, allowing the mass transfer between them with losses thermal minimum of transfer and eliminates the need for exchanger of external heat of recovery. P. Mavroudaki and C.B. Beggs [7] reviewed the application of desiccant dehumidification cooling systems in northern Europe and concluded that there was potential to utilize these technologies in the southern European countries.

Nomenclature			
C	Specific heat capacity (J/kg.°C)	Subscripts	
CC	Cooling Capacity (W)	Cool	Cool
COP	Coefficient of performance	El	Electrical
M	Flow rate (kg/s)	Exp	Experimental
Q	Rate of heat transfer (W)	Heat	Heat
W	Humidity ratio (g/kg)	S	Dry air
Greek letters		Sol	Solution
Θ	Temperature (°C)	Theo	Theoretical

In this paper we present the comparison between a theoretical simulation and experimental results of the influence of climatic parameters in the coefficient of performance and cooling capacity of a desiccant cooling system. The project aims at assessing the effectiveness of this kind of a heat recovery/desiccant cooling technologies under hot and humid climate.

2. Description system

The liquid desiccant system is designed to serve as open-cycle absorption system that can operate with low grade solar heat. The system consists of six components: a dehumidifier, a regenerator, two heat exchangers (water-to-solution), a heat exchanger (solution-to-solution), and air-to-air heat exchanger, a schematic description of a system is given in Figure 1.

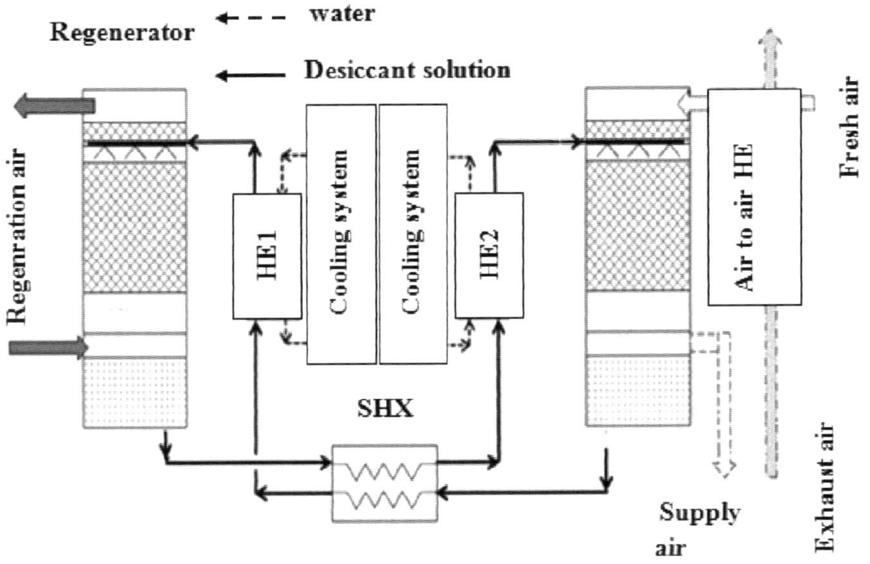

Figure 1. Description of liquid desiccant system.

When the fresh air enters the system, first it is cooled in the heat exchanger recovery (air-to-air heat exchanger), then it enters into the dehumidifier, it makes contact with the concentrated and cool liquid desiccant, where it is dehumidified and cooled.

The liquid desiccant cycle is started by pumping the solution at the bottom of the dehumidifier, passes through the solution to solution heat exchanger (SHX) to preheat it, then it is heated to attain regeneration temperature by solar energy, and finally sprayed in the regenerator to compress.

3. Theoretical analysis

Many researchers have developed mathematical models of the coupled heat and mass transfer processes in the dehumidifier or/and regenerator of most of models were solved numerically [4].
The output energy (cooling capacity CC) is energy reduction from the fresh air to supply air as shown in the following:

$$CC = m_{air} * C_s * \left(\theta_{freshair} - \theta_{\text{supplyair}}\right) \tag{1}$$

The temperature of cooled water leaving the cooling system (cooling tower) is assumed to be lower than the solution temperature .entering in the heat exchanger (HE2). The heat transfer rate (**Q_{cool}**) transferred from the solution to the cold water is calculated:

$$Q_{cool} = m_{sol4} * C_{sol} * \left(\theta_{sol4} - \theta_{sol5}\right) \tag{2}$$

The temperature of heated water leaving is assumed to be higher than the solution temperature entering in the heat exchanger (HE1). The heat transfer rate (**Q_{heat}**) transferred from the hot water to the solution is calculated:

$$Q_{heat} = m_{sol1} * C_{sol} * \left(\theta_{sol2} - \theta_{sol1}\right) \tag{3}$$

Electrical energy (Qel) consumed in this system, we take the electrical consumption of the example of Shuli Liu system [4], which require a small amount of energy with the maximal value of about 500W (There are 6 pumps (30W) and 3 fans (100W)).

We present the coefficient of performance (COP) to describe the working performance of the global system. It is defined as the ratio between the cooling capacity and the total primary energy consumption in the whole liquid desiccant system. We taken the same equation used by Shuli Liu to compare the theoretical results with her experimental results. The COP of the whole system is:

$$COP = \frac{CC}{Q_{cool} + Q_{el} + Q_{heat}} \tag{4}$$

4. Numerical results

We present a comparison between the experimental results of Shuli Liu and our theoretical simulation of the heat recovery desiccant cooling system. The input conditions to the theoretical simulation were the same input values of experimental testing conditions used by Shuli Liu [4].

4.1. Influence of fresh air humidity

The experimental and the theoretical simulation were effectuated to examine the air humidity's influence on the cooling capacity and the coefficient of performance of the whole system.
The average cooling water temperature was equal to 16°C and the flow rate was 0.23kg/s. the average heating water temperature was equal to 63°C with the flow rate of 0.15kg/s. With the same fresh and return air flow rates were 509m^3/h and the fresh air temperature equal to 30°C, the testing and theoretical results indicate that the fresh air humidity have an effects in the CC of the system.
The results of the experimental and theoretical CC of the system are shown in Figure 2. When the fresh air humidity increases from 11.43g/kg to 16.84g/kg, it provoked the increase of the experimental **CC_{exp}** from 2.83 to 4.14kW [4] and the theoretical **CCt_{heo}** from 3.24 to 3.8kW.
The highest of CC values due to the increase of enthalpy difference between the fresh and leaving air from the system. The average CC difference percentage between the theoretical and the experimental results is 10%. This average difference percentage is acceptable for the whole system and it is caused by some simulation errors.

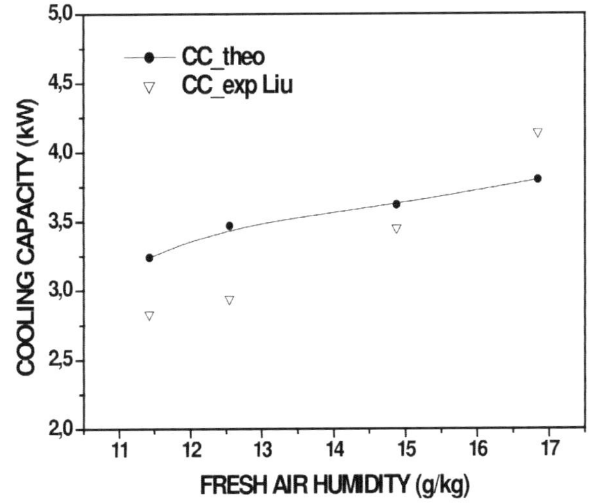

Figure 2. Influence of fresh air humidity on the cooling capacity.

4.2. *Influence of fresh air temperature*

The experimental and the theoretical simulation were carried to present the fresh air temperature's influence on the CC system. The average cooling and heating water temperature used in the experimental tests are 17°C and 61.5°C respectively. The fresh and return air flow rates were 509m^3/h and the cooling and heating water flow rates were 0.15kg/s and 0.16kg/s respectively [4].

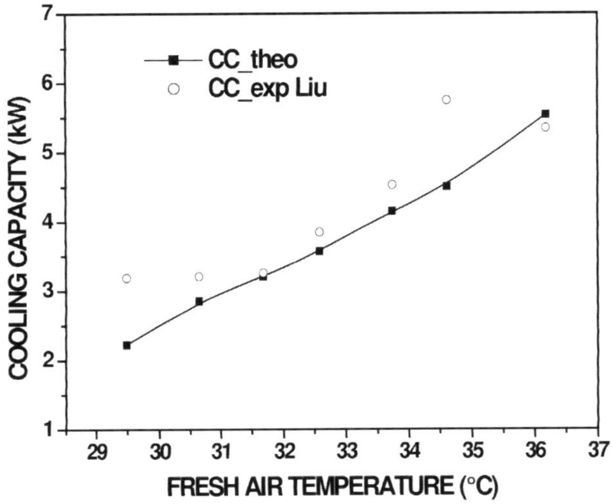

Figure 3. Influence of fresh air temperature on the cooling capacity.

In Figure 3, The experimental and theoretical results show that the CC of the system increase when the fresh air temperature increase, the highest fresh air temperature from 29.5°C to 36.18°C resulted the increase of the theoretical $\mathbf{CC_{theo}}$ of the system from 2.22kW to 5.53kW and the highest value of experimental $\mathbf{CC_{exp}}$ are about 5.35kW respectively are shown in Figure 3.
This increase is due to difference between the air enthalpy at the inlet and outlet of the dehumidifier in the whole system. According to these results, the liquid desiccant system is efficient in the hot and humid climates. The theoretical simulation results indicate the same increase in the experimental test with an average CC difference percentage is 14.5%.

4.3. Typical summer day results

The experimental and theoretical results show that this kind of system is very efficient in the hot and humid climate. Tunisia is situated on the Mediterranean coast of North Africa; it lies between latitudes 30° and 38°N, and longitudes 7° and 12°E [8]. It has a high level of solar power with an average direct irradiation which varies from 2 kWh/m^2 /d in the extreme north to 6 kWh/ m^2/d in the zones of the extreme south [9]. Tunisian climate characterized by a hot and humid climate in the summer. In this part we test the system under a summer Tunisian day.
Figure 4 presents the solar radiation and ambient (fresh) air temperature of a sunny day representative in the August month chosen from data file. In summer, it is expected that the temperature of the fresh humid air that enters the cooling system is higher than the inlet air temperature in the build. At 13h, the ambient air temperature is about 40°C with a humidity ratio equal to 23.9 g/kg$_{dry\ air}$ and a global solar radiation is more than 800 W/m².

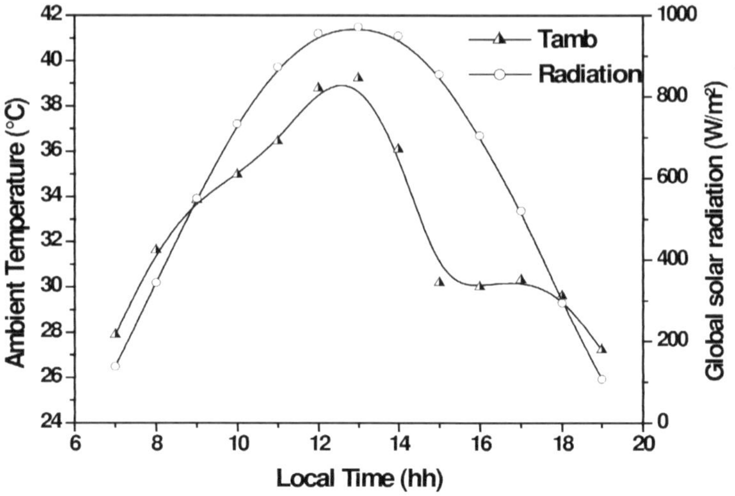

Figure 4. The solar radiation and ambient air temperature of a summer Tunisian day.

Figure 5 shows the air specific humidity at the inlet (fresh air) and outlet (supply air) of the cooling system under the indicated summer climatic day. The air specific humidity decreased between 40-65 %, this deference between inlet and outlet is proportional to moisture transfer during an air dehumidification process.

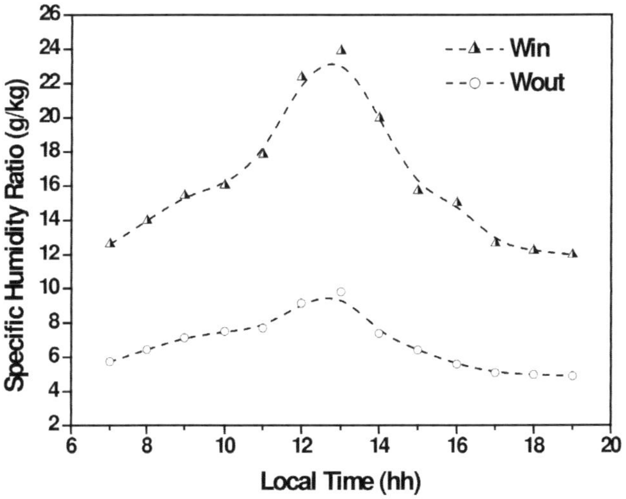

Figure 5. Air specific humidity at the inlet and outlet of the cooling system.

The air temperature at the inlet (fresh air) and outlet (supply air) of the cooling system under the indicated climatic conditions are shown in Figure 6. The supply air temperature is expected to lower than the summer comfort condition (25°C) due to the contact of the air stream with the cool and strong desiccant solution. This is confirmed with the experimental data [4].

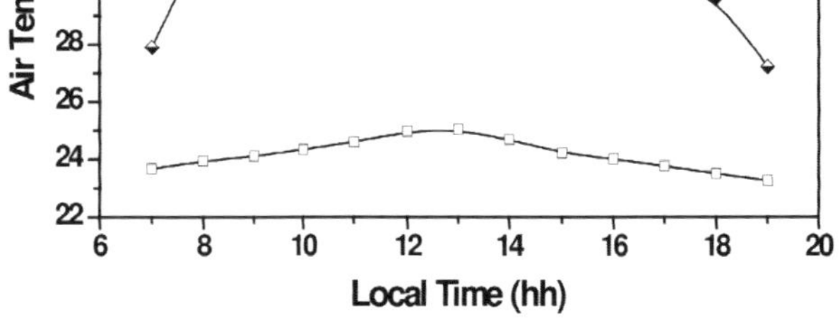

Figure 6. Air temperature at the inlet and outlet of the cooling system.

5. Conclusion

The objective of this study is to present a comparison between a theoretical simulation and experimental results of a desiccant cooling system which is efficient and can be very useful in the hot and humid climatic conditions. This comparison is to validate the simulation results of the proposed system. The simulation results show the influence of the climatic parameters (Fresh air temperature and Fresh air humidity) in the cooling capacity (CC) with acceptable percentage deference between the experimental and theoretical results. We test the system under a typical summer day of a hot and humid place (Tunis, Tunisia) and we conclude that the system can be a sustainable solution to operate in the Tunisian climate and the theoretical simulation should be followed by an experimental work under real conditions to prove the theoretical results and to improve the economic side of system in terms of energy efficiency.

References

[1] Ma Q, Wang R Z, Dai Y J, Zhai X Q 2006 *Energ. Build.* **38** 447-453
[2] Hellmann H M and Grossman G 1995 *Int. J. Refrig* **18** 177-189
[3] Mei L, Dai Y J 2008 *Renew.and Sust. Ener. Rev.* **12** 662-689
[4] Shuli Liu 2008 Thesis "A Novel Heat Recovery/Desiccant Cooling System" University of Nottingham.
[5] Gommed K and Grossman G 2012 *Int. J. of Refrig.* **35** 676-684
[6] Gommed K and Grossman G 2007 *Sol. Ener.* **81** 131-138
[7] Mavroudaki P, Beggs CB, Sleigh P A and Halliday S P 2002 *Appl. Ther. Eng.* **22** 1129-1140
[8] National Agency for Energy Conservation Tunisia (NAEC), http://www.anme.nat.tn.
[9] United Nations (UN), Office for Africa of north, the renewable energy sector for North Africa situation current and perspective Economic commission on Africa (ECA). p32, September2012.

Non-catalytic alcoholysis process for production of biodiesel fuel by using bubble column reactor

S Hagiwara[1], H Nabetani[1] and M Nakajima[2]

[1] National Food Research Institute, NARO, Tsukuba, Ibaraki 305-8642, Japan.
[2]Alliance for Research on North Africa (ARENA), University of Tsukuba. Tsukuba, Ibaraki 305-8572, Japan.

E-mail: shoji@affrc.go.jp

Abstract. Biodiesel fuel is a replacement for diesel as a fuel produced from biomass resources. It is usually defined as a fatty acid methyl ester (FAME) derived from vegetable oil or animal fat. In European countries, such as Germany and France, biodiesel fuel is commercially produced mainly from rapeseed oil, whereas in the United States and Argentina, soybean oil is more frequently used. In many other countries such as Japan and countries in Southeast Asia, lipids that cannot be used as a food source could be more suitable materials for the production of biodiesel fuel because its production from edible oils could result in an increase in the price of edible oils, thereby increasing the cost of some foodstuffs. Therefore, used edible oil, lipids contained in waste effluent from the oil milling process, by-products from oil refining process and crude oils from industrial crops such as jatropha could be more promising materials in these countries. The materials available in Japan and Southeast Asia for the production of biodiesel fuel have common characteristics; they contain considerable amount of impurities and are high in free fatty acids (FFA). Superheated methanol vapor (SMV) reactor might be a promising method for biodiesel fuel production utilizing oil feedstock containing FFA such as waste vegetable oil and crude vegetable oil. In the conventional method using alkaline catalyst, FFA contained in waste vegetable oil is known to react with alkaline catalyst such as NaOH and KOH generating saponification products and to inactivate it. Therefore, the FFA needs to be removed from the feedstock prior to the reaction. Removal of the alkaline catalyst after the reaction is also required. In the case of the SMV reactor, the processes for removing FFA prior to the reaction and catalyst after the reaction can be omitted because it requires no catalyst. Nevertheless, detailed study on the productivity of biodiesel fuel produced from waste vegetable oils and other non-edible lipids by use of the SMV reactor has not been examined yet. Therefore, this study aims to investigate the productivity of biodiesel produced from waste vegetable oils using the SMV reactor. Biodiesel fuel is a replacement for diesel as a fuel produced from biomass resources. It is generally produced as a FAME derived from vegetable oil by using alkaline catalyzed alcoholysis process. This alkaline method requires deacidification process prior to the reaction process and the alkaline catalyst removal process after the reaction. Those process increases the total cost of biodiesel fuel production. In order to solve the problems in the conventional alkaline catalyzed alcoholysis process, the authors proposed a non-catalytic alcoholysis process called the Superheated Methanol Vapor (SMV) method with bubble column reactor. So, this study aims to investigate the productivity of biodiesel produced from vegetable oils and other lipids using the SMV method with bubble column reactor.

Journal of Physics: Conference Series **596** (2015) 012017 doi:10.1088/1742-6596/596/1/012017

1. Introduction

Biodiesel fuel is a replacement for diesel (fossil fuel) as a fuel produced from biomass resources. Used edible oil, lipids contained in waste effluent from the oil milling process, by-products from oil refining process and crude oils from industrial crops such as Jatropha could be more promising materials for production of biodiesel fuel in Japan and Southeast Asia. Those materials contain considerable amount of impurities and are high in free fatty acids (FFA).

Non-catalytic Superheated methanol vapor (SMV) method using bubble column reactor might be a promising method for biodiesel fuel production utilizing oil feedstock containing FFA such as used vegetable oil and crude vegetable oil [1, 2]. In the conventional method using alkaline catalyst, FFA contained in used vegetable oil is known to react with alkaline catalyst such as NaOH and KOH generating saponification products and to inactivate it. Therefore, the FFA needs to be removed from the feedstock prior to the reaction. Removal of the alkaline catalyst after the reaction is also required. In the case of the SMV reactor, the processes for removing FFA prior to the reaction and catalyst after the reaction can be omitted because it requires no catalyst [3, 4]. Nevertheless, detailed study on the productivity of biodiesel fuel produced from used vegetable oils by using SMV method has not been examined yet. Therefore, this study aims to investigate the productivity of biodiesel produced from used vegetable oils using SMV method with bubble column reactor.

2. Promising materials for production of biodiesel fuel

Biodiesel fuel is a replacement for diesel as a fuel produced from biomass resources. It is usually defined as a fatty acid methyl ester (FAME) derived from vegetable oil or animal fat.

In European countries, such as Germany and France, biodiesel fuel is commercially produced mainly from rapeseed oil, whereas in the United States and Argentina, soybean oil is more frequently used.

In many other countries such as Japan and countries in Southeast Asia, lipids that cannot be used as a food source could be more suitable materials for the production of biodiesel fuel because its production from edible oils could result in an increase in the price of edible oils, thereby increasing the cost of some foodstuffs. Therefore, used edible oil, lipids contained in waste effluent from the oil milling process, by-products from oil refining process and crude oils from industrial crops such as jatropha could be more promising materials for use in these countries.

In Japan, it is not feasible to use virgin vegetable oil as a raw material for producing biodiesel fuel because Japan has to import a large quantity of the edible oil from foreign countries. However, it could be possible to utilize used edible oil. In Japan, about 9000 kL/y of FAME is produced from used edible oil and utilized as biodiesel fuel.

Waste effluent from oil milling processes and by-products from oil refining processes contain substantial amounts of lipids. For example, FFA are obtained from the refining process as by-products. The FFA account for 5-10 % of crude oil. These materials are present already and need not be collected. Therefore, if these materials are converted into FAME at a low cost, they could provide a useful source of biodiesel fuel.

Jatropha (*Jatropha curcas L.*) is an oil crop, which can grow in marginal land such as the eastern part of Indonesia where the amount of rainfall is less than 400 mm/y (Figure 1). The oil content of jatropha seeds is 30-40 %. Oil productivity of the jatropha plant is considered higher than that of soybean or rapeseed. The first harvest can be obtained within eight months of planting. Then, about 20 kg of seed can be harvested twice a year from one plant. Plant life is about 50 years. However, jatropha oil contains toxins and is unsuitable for human consumption. Therefore, utilization of jatropha oil in the production of biodiesel does not affect the supply and price of edible oil. If crude jatropha oil can be converted into FAME at a low cost, it will be a promising material for biodiesel fuel. Agriculture could therefore be expanded into marginal land by introducing jatropha plants, which will benefit the residing communities.

As mentioned above, the materials available in Japan and Southeast Asia for the production of biodiesel fuel have common characteristics; they contain considerable amount of impurities and are high in FFA.

Figure 1. Jatropha plants 6 months after planting (A), fruits (B), seeds (C), and crude oil obtained by screw press (D)..

3. Problems with the conventional alkaline catalyzed alcoholysis reaction process for the production of biodiesel fuel

In the alcoholysis reaction for the production of biodiesel fuel from vegetable oil, triglyceride, which is the main component of vegetable oil, is reacted with methanol to form FAME and glycerol. FAME is used as a biodiesel fuel. In the conventional process, alkaline catalysts such as NaOH and KOH are used to promote the reaction. In the FAME production process using the alkaline catalyst method, the catalyst has to be removed from the products after the reaction. Otherwise, glycerol which is a by-product of the process cannot be utilized directly in other industries. However, the process of removing the alkaline catalyst increases the total cost of biodiesel fuel production.

In the case of the alkaline catalyzed alcoholysis reaction process, FFA contained in raw materials have to be removed prior to the reaction in order to maintain the activity of the catalyst. However, the removal of FFA reduces the yield of the process.

4. Principle and advantages of a novel non-catalytic alcoholysis reaction process called superheated methanol vapor method using bubble column reactor

In order to solve the problems in the conventional alkaline catalyzed alcoholysis reaction process for the production of biodiesel fuel, the authors proposed a non-catalytic alcoholysis reaction process called the superheated methanol vapor method using bubble column reactor.

Figure 2 shows the schematic flow diagram of a reactor for the superheated methanol vapor method using bubble column reactor [1, 2]. Superheated methanol vapor is continuously bubbled into the oil in the reactor vessel and reacted with triglycerides to form FAME and glycerol. The FAME and glycerol formed flows out of the reactor together with unreacted methanol vapor and is collected using a condenser. Condensed FAME is separated from glycerol by sedimentation and used as biodiesel fuel. Unreacted methanol vapor is sent back to the reactor vessel and reused. No catalyst is used in this process.

In a system with the superheated methanol vapor method using bubble column reactor, the purification process to remove the catalyst after the reaction is not required. Therefore, configuration of the total system can be simplified and the by-product, glycerol, can be directly utilized in other industries. In this way, the total cost of biodiesel fuel production will be reduced. Furthermore, triglycerides as well as FFA could be converted into FAME in this method. Therefore, the neutralization process for the removal of FFA is not required prior to the reaction process. This eventually improves the yield of the total system.

Figure 3 shows the effects of material composition on the outflow rate of FAME in the superheated methanol vapor method. This result shows that not only triglyceride but also FFA can be converted into FAME using this method. The outflow rate is several times higher than that with triglycerides [3, 4]. Furthermore, the result shows that addition of FFA enhances the conversion of triglyceride to FAME. This result implies that this method is suitable for materials available in Japan and Southeast Asia, which are rich in FFA.

In order to confirm the applicability of the superheated methanol vapor method to non-edible crude oils, crude jatropha oil was processed with the method. Crude jatropha oil showed comparable FAME productivity to edible oil such as rapeseed and soybean oils.

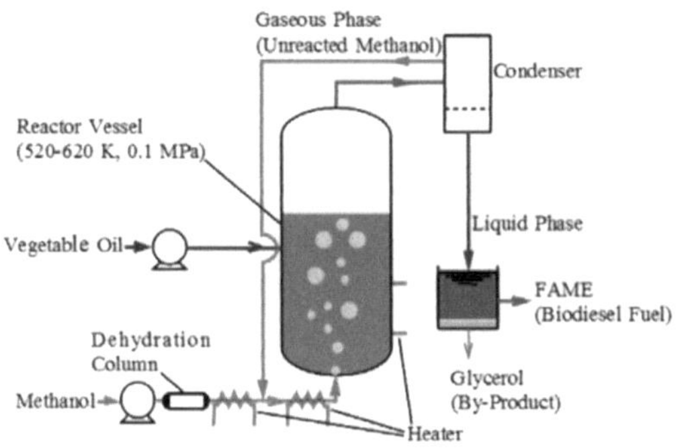

Figure 2. Schematic flow diagram of a reactor based on the superheated methanol vapor bubble method (FAME: fatty acid methyl ester).

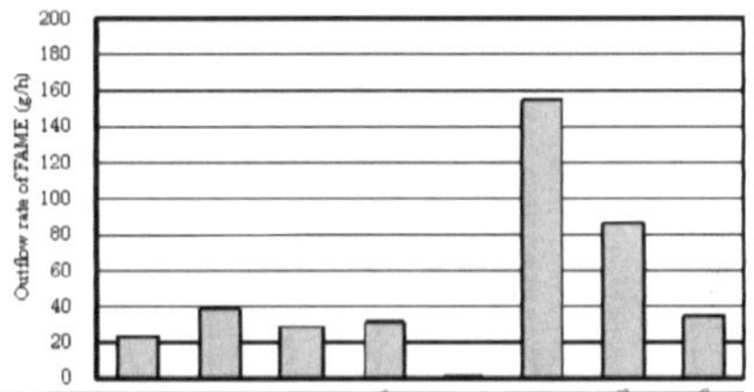

Sample		Fresh Oil (FO)	Used Oil	DG	FO:DG =50:50	MG	Oleic Acid (OA)	FO:OA =50:50	FO:OA =98:2
Composition	TG	98.7	90.5	13.5	56.0	-	-	49.3	96.4
	DG	1.0	5.0	82.5	41.8	-	-	0.5	1.0
	MG	0.4	4.0	4.0	2.2	100.0	-	0.2	0.4
	FFA	-	0.5	-	-	-	100.0	50.0	2.2
Outflow rate of FAME (g/h)		23.1	38.5	28.8	31.0	1.2	154.4	85.9	34.8

Figure 3. Effects of impurities contained in waste edible oils on outflow rate of fatty acid methyl ester (FAME) from a reactor based on the superheated methanol vapor bubble method (TG: triglyceride. DG: diglyceride. MG: monoglyceride. FFA: free fatty acid.).

5. Economical efficiency of a non-catalytic alcoholysis reaction process using superheated methanol vapor method with bubble column reactor

Eight years ago, the authors' group designed and constructed a demonstration plant based on superheated methanol vapor method (Figure 4). This plant showed 425 L/d productivity of FAME from used frying oil. And energy consumed in each unit of the demonstration plant was measured (electrical energy and thermal energy). Then, based on the energy consumption data obtained with the demonstration plant, production cost required with a practical scale plant (designed productivity: 6000 kL/y) was calculated (Table 1).

The cost required by the practical scale plant with the superheated methanol vapor method was estimated to be 40.6 yen/L (about 40 cent/L) while the cost required by a plant with the alkaline catalyst method was 62.5 yen/L (about 62 cent/L). The estimated cost includes depreciation cost, cost of repairing, labor cost, methanol cost and energy cost (heat and electricity). Most of the energy consumed by the plant was thermal energy and the plant could be automatically controlled. Therefore, required cost will be further decreased by installing the plant next to an incineration facility because thermal energy can be supplied from the facility and the labor cost can also be supported by the facility.

Figure 4. Demonstration plant based on superheated methanol vapor method using bubble column reactor (Designed Productivity: 400 L/d)

Table 1. Estimated cost for production of biodiesel fuel by using superheated methanol vapor method with bubble column reactor

		Demonstration Plant (146 kL/y)	Practical Scale Plant (6 000 kL/y) Basic Case	Practical Scale Plant (6 000 kL/y) Combination with an incineration facility	Alkaline Catalyst Method (1500 kL/y)
Depreciation Cost	[yen/kg]	39.2	8.6	8.6	36.1
Cost of Repairing	[yen/kg]	0.9	0.2	0.2	0.8
Labor Cost	[yen/kg]	313.9	7.3	-	7.3
Methanol Cost	[yen/kg]	18.0	10	10.0	25.2
Energy Cost (heat)	[yen/kg]	82.7	13.6	-	1.2
Energy Cost (electricity)	[yen/kg]	28.9	9.7	6.4	1.5
Cost of Catalyst	[yen/kg]	-	-	-	4.1
Total	[yen/kg]	483.6	49.5	28.5	76.1
	[yen/L]	396.5	40.6	23.4	62.5

6. Conclusions

Economical efficiency of a non-catalytic alcoholysis process using bubble column reactor for production of biodiesel fuel was demonstrated in this paper. The authors hope that the results obtained in this research work will reduce the cost involved in producing biodiesel fuel and contribute to the prevention of global warming, reduction in consumption of fossil fuels and the utilization of rural areas in Japan as well as in many other countries.

Acknowledgements

Research on superheated methanol vapor method with bubble column reactor was supported by a grant from the MAFF, Japan (Rural Biomass Research Project, BEC-BF010) and JSPS KAKENHI Grant Numbers 00353970.

References

[1] Yamazaki R, Iwamoto S, Nabetani H, Osakada K, Miyawaki O and Sagara Y 2007 *Jpn. J. Food Eng.* **8** 11

[2] Joelianingsih, Maeda H, Hagiwara S, Nabetani H, Sagara Y, Soerawidjaya T H, Tambunan A H and Abdullah K 2008 *Renew. Energ.* **33** 1629

[3] Joelianingsih, Nabetani H., Hagiwara S, Sagara Y, Soerawidjaya T H, Tambunan A H and Abdullah K 2007 *J. Chem. Engi. Jpn.* **40** 780

[4] Joelianingsih, Nabetani H, Sagara Y, Tambunan A H and Abdullah K 2012 *Fuel* **96(6)** 595

Optical study of solar tower power plants

F Eddhibi, M Ben Amara, M Balghouthi, and A Guizani
Thermal Process Laboratory, Research and Technology center of Energy P.B N°95 2050, Hammam Lif – Tunisia.

E-mail: eddhibifethia@yahoo.fr

Abstract. The central receiver technology for electricity generation consists of concentrating solar radiation coming from the solar tracker field into a central receiver surface located on the top of the tower. The heliostat field is constituted of a big number of reflective mirrors; each heliostat tracks the sun individually and reflects the sunlight to a focal point. Therefore, the heliostat should be positioned with high precision in order to minimize optical losses. In the current work, a mathematical model for the analysis of the optical efficiency of solar tower field power plant is proposed. The impact of the different factors which influence the optical efficiency is analyzed. These parameters are mainly, the shading and blocking losses, the cosine effect, the atmospheric attenuation and the spillage losses. A new method for the calculation of blocking and shadowing efficiency is introduced and validated by open literature.

1. Introduction

Central receiver system consists of a high sun-tracking mirror number which concentrates solar radiation at a focal point located on the top of a high tower [1]. The heliostat field is the central subsystem in the solar tower power plant; it contributes about 50% to the total cost [2] and causes about 47% of the annual energy losses [3]. Several studies dealt with the optimization and the design of heliostat field layout [4]. Owing to the high heliostat number, the calculation of the optical losses is an iterative process [5]. Two categories of codes have been developed; the first one is dedicated only for the optical analysis of a solar concentrating system (MIRVAL, FIATLUX and SOLTRACE). The second category (DELSOL, UHC and HFLCAL) is designed to the total optimization of the concentrating system [5,6].

Many types of losses affect the heliostat field optical efficiency, namely cosine loss, losses due to the shadowing, blocking, atmospheric attenuation and spillage losses. In this current work, a mathematical model for the analysis of the optical efficiency of solar tower power plant is proposed. To validate the new proposed method, a detailed calculation of the PS10 solar field optical efficiency is made based on optical losses with wind effects neglected [6]. Shadowing and blocking effects which are the key parameters on heliostat efficiency are the main variables targeted in this work.

2. Model description

Shading and blocking efficiency is the most expensive efficiency evaluation parameters. Several calculation methods are available in open literature such the one defined by Lipps [7], this method is called the single vertex. Another method is used by HELIOS [8]: Who used the cone optics method to evaluate the solar flux density on the receiver.

According to the number of heliostats, the simulation time is too large. In order to minimize the running time, a new method is developed in this paper based on the minimization of the numbers of equations and simplification of the used methods.

2.1. PS10 solar power plant

Table 1. PS10 parameters [10]

Location	Latitude [°]	37.44
	Longitude [°]	-6.25
Heliostat	Width [m]	12.84
	Height [m]	9.45
	Pillar height [m]	5.17
Tower	Height [m]	115
	Length [m]	18
	Width [m]	8
Receiver	Height of receiver bottom [m]	100.5
	Inclination angle [°]	11.5
	Width [m]	13.78
	Height [m]	12

2.2. Sun position

The global coordinate system includes all geographical coordinates: Altitude, Longitude, Latitude and time. With reference to these coordinates, solar position is defined by the solar azimuth, and the solar zenith [10].

3. Heliostat field efficiency

The instantaneous heliostat field efficiency is the product of the instantaneous optical loss factors: cosine losses (η_{cos}), the atmospheric attenuation (η_{atm}), heliostat reflectivity (η_{ref}), shadowing and blocking losses (η_{sb}), spillage losses ($\eta_{spillage}$) [11].

$$\eta = \eta_{cos} \times \eta_{atm} \times \eta_{ref} \times \eta_{sb} \times \eta_{spillage} \tag{1}$$

3.1. Cosine losses

In order to guarantee an accurate tracking of the sun path during a day, each heliostat has to change its orientation according to incident angle. The cosine efficiency is calculated as the dot product of the incident sun ray direction and the normal of the mirror surface [12].

$$\eta_{cos} = \vec{\imath} \times \vec{n} \tag{2}$$

3.2. Atmospheric attenuation

During its path from mirror to the receiver, reflected rays are affected by the atmospheric attenuation. Losses due to the atmospheric attenuation are function of the distance between heliostat and the receiver located on the top of the tower. Also, it depends on some weather conditions i.e. the visibility. It can be calculated as [10]:

$$\eta_{atm} = \begin{cases} 0.99321 - 0.0001176 \times \text{dist} + 1.97 \times 10^{-8} \times \text{dist}^2 & \text{dist} \leq 1000\text{m} \\ \exp(-0.0001106 \times \text{dist}) & \text{dist} > 1000\text{m} \end{cases} \tag{3}$$

3.3. Mirror reflectivity
It is the quality of reflective surface; it depends on degradation and cleanliness. In this study it is assumed to be constant equal to 0.88 [10].

3.4. Spillage loss
A portion of reflected radiation losses its path and cannot hit the receiver. This is caused by tracking accuracy and mirror quality. This phenomenon is called Spillage.

3.5. Blocking/Shading loss
Blocking means that the heliostat located behind its neighbour cannot reflect its entire surface to the receiver, a portion of the reflected rays is blocked by the back side of the front heliostat. Shading is similar to the blocking, but it affects incident rays, heliostat placed in front of its neighbour shads the heliostat behind. A detailed description of the shadowing and blocking method calculation is presented on the next section.

4. A new method for the blocking and shading losses

4.1. Local coordinate system
The 3D coordinate system is defined as follows: the x axis goes from west to east, y axis goes from south to north and the z goes from the ground to the sky, it is confounded with the tower.

4.2. Heliostat coordinates
Heliostat position is defined by its Cartesian coordinates (x_i, y_i, z_i). The distance between a heliostat and the tower is calculated by equation (5). Each heliostat admits its own latitude and its own angle of inclination.

$$hel_{position} = \sqrt{x_i^2 + y_i^2 + z_i^2} \tag{4}$$

The altitude of a heliostat (i) is Al_i, it is defined as the angle between the horizontal planes passing through the center of the heliostat and the reflected ray.

$$Al_i = \tan^{-1}\left(\frac{H_{tower} - z_i}{hel_{position}}\right) \tag{5}$$

The angle of inclination (B) is:

$$B = \begin{cases} y_i > 0, \; B = \cos^{-1}\left(\frac{n_3}{n_{module}}\right) \\ y_i < 0, \; B = 90 - \cos^{-1}\left(\frac{n_3}{n_{module}}\right) \end{cases} \tag{6}$$

4.3. Blocking/shadowing
The calculation procedure for the blocking and shading efficiency can be limited only to three steps:

- Select of the heliostat neighbour: In this section, only the x and y coordinates are considered. The k_d tree method [13,14] is used to find the heliostat neighbour: it is a binary tree which aims to find the nearest neighbour.
- Reduction of the computational complexity: boundary conditions: the k_d tree outputs are 10 neighbours for each heliostat, which means a total of (624*10) for the entire field. Therefore, we need to reduce this computational complexity. The same boundary conditions declared by M.Ewert et al. [10] are used. The number of heliostat which shad/block the actual heliostat will be function of the sun position.

- In this step, a new referential is defined; each heliostat id defined as a rectangle with a center O (0, 0, 0), therefore we can define the corner as follow:

A (L/2; W/2; 0)
B (-L/2; W/2; 0)
C (-L/2; -W/2; 0)
D (L/2; -W/2; 0)

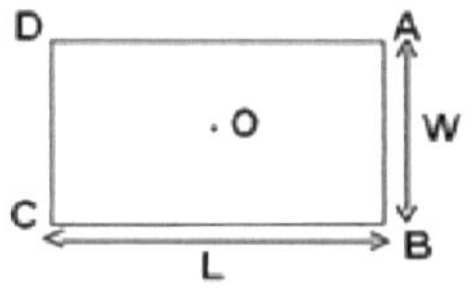

In the case of shading, each corner will be projected over the opposite sun vector. Then we have to check if the projected point is located or not on the neighbour surface. Each surface is defined by a point (center in this case) and a normal vector. In our case the actual heliostat is defined by its center (x0, y0, z0) and its normal vector. If the projected point belongs to the actual heliostat, it should follow the formula:

$$\begin{pmatrix} x_0 - x_{proj} \\ y_0 - y_{proj} \\ z_0 - z_{proj} \end{pmatrix} \times \begin{pmatrix} n_1 \\ n_2 \\ n_3 \end{pmatrix} = \begin{pmatrix} 0 \\ 0 \\ 0 \end{pmatrix} \tag{7}$$

Finally, the intersection between the actual heliostat surface and the projected surface defines the shaded area. The shading efficiency is defined as:

$$\eta_{shd} = 1 - \frac{Shaded\ surface}{heliostat\ surface} \tag{8}$$

The difference between the blocking and shading is that the blocked area is calculated on the neighbouring heliostat and the shaded area is a part of the heliostat itself.

5. Results and discussion

5.1. Validation

The new mathematical method of optical efficiency calculation is validated based on PS10 data. PS10 solar field heliostat coordinates was generated by the Sandia code WINDELSOL1.0. Blocking and shading are calculated by projecting the heliostat neighbour along the sun direction or the reflected ray direction onto the plane of the considered heliostat [16]. In this case ten neighbours are considered in the simulation. The individual efficiency factors are detailed in the table below:

Table 2. Individual efficiency factor of PS10

Efficiency	Value
η_{cos}	0.8283
η_{atm}	0.9498
η_{ref}	0.8800
η_{sb}	0.9255
$\eta_{spillage}$	0.9926

Table 3. PS10 Field efficiency validation

	Ref [15]	Ref[11]	New code
Cosine efficiency	0.824	0.8283	0.8241
Atmospheric attenuation	0.950	0.9498	0.9500
Shadowing and blocking efficiency	0.930	0.9255	0.9218

According to table 2, the results are in good agreement with those published by Xiudong Wei et al. [15] and Corey J. N et al. [11].

5.2. Cosine efficiency and attenuation efficiency

After validation we achieve a parametric study to evaluate the effect of key parameters on the efficiency of the field. The PS10 parameters are considered constant in all the simulation. The simulation time is chosen, in this case, as: 1st march at 12:00 local time (Seville). The sun elevation is 39.36 and the sun azimuth is 127.30. We report in the next figure the variation of the efficiency. The color degradation indicates the range of losses: the blue color indicates the range less than 0.6 and the red one is the maximum value of efficiency greater than 0.80. The figure shows that the Eastern part of the heliostat field is less affected by the cosine losses than the western part. The individual cosine efficiency is in the range of 0.8241.

Compared to the value in the table above the error of calculation is less than 0.01%. The second figure shows the attenuation efficiency function of the distance between the heliostat and the receiver. In the case of heliostat positioned in the first ranges, the efficiency is more important (> 0.96). The individual attenuation efficiency factor is in the range of the value declared by Kistler [16].

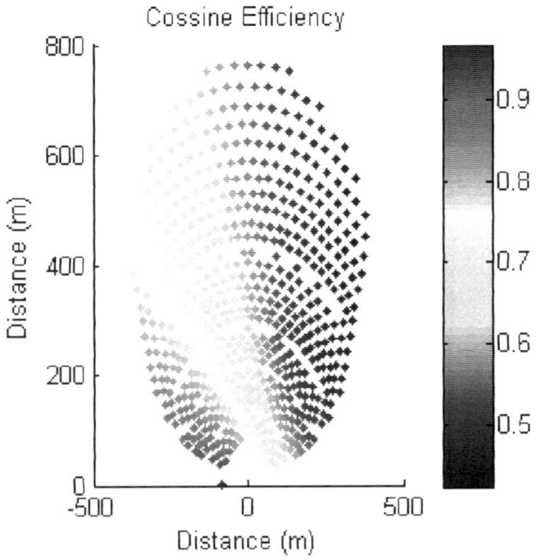

Figure 1. Cosine efficiency.

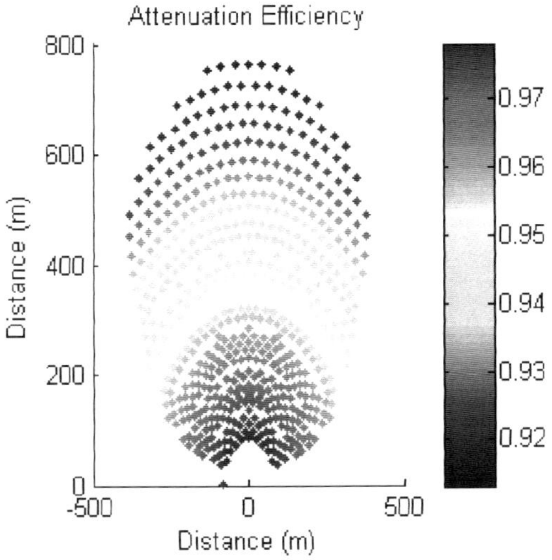

Figure 2. Attenuation efficiency

5.3. Shadowing and blocking efficiency
As mentioned above, the number of heliostats which shad/block the heliostat subject of our interest varies with the sun position. At low sun elevation the shadow is longer and a heliostat can affect more than one neighbour. The results show that it is also function of distance between two neighbours.
In the high density section of the heliostat field where the heliostats are closest (the first rows), shadowing and blocking efficiency is greater than 0.9825 and it decreases by moving away from the tower (low density region), when it reaches 0.6847.

6. Conclusion
Power production using tower technology needs a heliostat field which consists of a big number of reflective mirrors, each one moves independently of the others, and reflects the sunlight to a focal point located on the top of a tower. The motion of the heliostats generates: shading and blocking losses which are considered to be the most important optical losses. Several studies dealt with the optimization of the optical efficiency. In this current work, a new mathematical model for the analysis of the optical efficiency is proposed. The PS10 tower power plant system (Seville, Spain) is chosen as a demonstrative prototype. The model is detailed herein and validated with the open literature. Our results are in good agreement with those published by Xiudong Wei and Corey J. Noone. A parametric study to evaluate the effect of key parameters on the efficiency of the field is achieved and showed that the development of the new method can improve the existed method used to calculate heliostat field efficiency.
Once the instantaneous efficiency is calculated, a year-long efficiency calculation is needed in order to evaluate the average annual heliostat efficiency.

Acknowledgments:
The author thanks Dipl-Math, Pascal Richter, center of computational engineering science RWTH Aachen University Germany, for the valuable discussion with him which we much appreciated.
The research was established in collaboration with Sun to Market Solutions, the author is very grateful to S2M team, especially the head of the society and Ing. Almudena Ruano San José for their collaboration.

7. References
[1] Gregory J K. 1998 *Sol. En.* **62** 51-61
[2] Gregory J K, Scott A J, Matthew W D, David G, Robert T, Roger D and Ron L and al. 2007 *Heliostat Cost Reduction Study* (Sandia National laboratories)
[3] http://www.powerfromthesun.net/Book/chapter10/chapter10.html
[4] Xiudong W, Zhenwu L, Weixing Y and Zhifeng W. 2010. *Sol. Energ.* **84** 685-690
[5] Garcia P, Ferriere A, Bezian J J. 2008 *Sol. En.* **82** 189-197
[6] Johannes S, Bryan O and Daniel N *enerMENA CSP teaching Materials Chapter 8: Solar Tower Technology (*German Aerospace Center (DLR))
[7] Lipps F W. Vant-Hull LL. 1978 *Sol. En.* **20** 505-516
[8] Biggs F and Vittitoe CN. 1977 *HELIOS: a computational model for solar concentrators* (Labs, Albuquerque, NM, US-USSR workshop on Solar Energy Appl., Moscow)
[9] Romero M. and Gonzãlez-Aguilar J. 2014 *WIREs En Env* **3** 42–59
[10] M.Ewert, O.Navarro Fuentes Modeling and simulation of a solar tower power plant http://www.mathcces.rwth-aachen.de/_media/5people/frank/solartower.pdf
[11] Corey J.N, Manuel T and Alexander M 2012 *Sol. En.* **86** 792-803
[12] Erminia L, Bruno D.A 2011 *En* **36** 4828-37
[13] http://en.wikipedia.org/wiki/K-d_tree
[14] Jerome H.F, Jon L. B, Raphael A. F 1976 "An Algorithm for finding Best matches in Logarithmic Expected time" SIAC-PUB-1549 (Rev.) February
[15] Xiudong W, Zhenwu L, Zhifeng W, Weixing Y, Hongxing Z and Zhihao Y 2010 *Renew. En.* **35** 1970-75

[16] Kistler B L.1986. *A User's Manual for DELSOL3: A Computer Code for Calculating the optical performance and Optimal System Design for Solar Thermal Central Receiver Plants* (Sandia National laboratories)

Nomenclature:

α	Solar Azimuth, [°]
γ	Solar Zenith, [°]
η	Optical efficiency
η_{cos}	Cosine efficiency
η_{atm}	Atmospheric attenuation
$\eta_{spillage}$	Spillage losses
η_{ref}	Heliostat reflectivity
η_{sb}	Shadow and blocking losses
η_{shd}	Shadowing efficiency
$\vec{n}$	Normal vector to the heliostat surface
$\vec{\imath}$	Incident vector
$\vec{r}$	Reflected vector
x_i, y_i, z_i	Heliostat coordinates
$x_{proj}, y_{proj}, z_{proj}$	Projected point coordinates
x_0, y_0, z_0	Center coordinates
H_{tower}	Tower height [m]
$hel_{position}$	Heliostat position [m]
n_{module}	Module of the n vector
L	Mirror length [m]
W	Mirror width [m]

Study of $Cu_2ZnSn(S,Se)_4$ Thin Films for Solar Cell Application

Muhammad Monirul Islam[1,2,*], Mohammad Abdul Halim[2], Chon Joy[2], Xianjia Luo[2], Takeaki Sakurai[2], Noriyuki Sakai[3], Takuya Kato[3], Hiroki Sugimoto[3], Hitoshi Tampo[4], Hajime Shibata[4], Shigeru Niki[4], Akimoto Katsuhiro[1,2]

[1]Graduate School of Pure and Applied Sciences, Alliance for Research on North Africa (ARENA), University of Tsukuba, Japan.
[2]Institute of Applied Physics, University of Tsukuba, Japan.
[3] Energy Solution Business Center, Showa Shell Sekiyu K.K., Japan.
[4]National Institute of Advanced Industrial Science and Technology (AIST), Japan.

E-mail: islam.monir.ke@u.tsukuba.ac.jp

Abstract. $Cu_2ZnSn(S,Se)_4$ (CZTSSe) thin films with various Cu/Sn ratio in the films have been investigated to study the effect of compositional variation over the electrical, optical, and structural properties of the film. Surface morphology and grain size were found to be significantly influenced by the Cu/Sn ratio in the films and grain size was found better for the samples with moderate Cu/Sn ratio of 1.75. Irrespective of the growth condition and compositional variation, all the CZTSSe crystals show that grains are oriented along (112) direction as evident from the room temperature XRD data. Dark current-voltage (*I-V*) curve reveals that that sample with Cu/Sn = 1.75 exhibits lowest leakage current, while sample with Cu/Sn = 1.85 has the highest leakage current along with larger ideality factor indicating larger recombination centers in this film. Series resistance was also found to be higher in the sample with higher Cu-content. An anomaly in the optical band-gap has been explained with the presences of impurity phases and compositional inhomogeneities in the CZTSSe materials.

1. Introduction

In order to fulfil the global increasing energy-demand using technologies which are compatible with the realization of green environment, recently various alternative sources of electricity have been emerged including photovoltaic (PV) technology; more specifically solar cells. Immediate challenge of the PV technology is the cost of electricity generation. Therefore, research and development of solar cells with earth-abundant non-toxic absorber material along with low production cost should be of highly prioritized. Quaternary compound $Cu_2ZnSn(S,Se)_4$ (CZTSSe) is a promising material for absorber layer in the solar cells [1-3]. CZTSSe utilizes Zn and Sn in replacement of scarce material, Ga and In in the $Cu(In,Ga)Se_2$. Thus, CZTSSe becomes a potential alternative to chalcogenide-based thin film solar cells including $Cu(In,Ga)Se_2$ and CdTe mainly because of abundance and non-toxicity of the constituent elements. Band gap of the $Cu_2ZnSnSe_4$ is 1.04 eV [4], while it becomes 1.5 eV [5] for its Cu_2ZnSnS_4 counterpart. Thus, band-gap of the $Cu_2ZnSn(S,Se)_4$ can be controlled by varying the Se/S ratio in the material. In case of the single junction solar cell, the ideal band-gap of the absorber layer to achieve the highest conversion efficiency under AM 1.5 sunlight is speculated

theoretically to be around 1.4 eV [6]. However, laboratory-scale efficiency of $Cu_2ZnSn(S,Se)_4$ solar cell so far has been reported as high as ~12 % [7], while pure Cu_2ZnSnS_4 shows efficiency around 8 % [8]. Moreover, efficiency is still behind the $Cu(In,Ga)Se_2$ counterpart which is around 21 % [9].
A complete understanding of the growth mechanism, growth condition and their correlation with the material properties, and their impact over the device performances must be the key issue to achieve efficiency beyond the current limit. The electrical, optical and micro-structural properties of the CZTSSe films are dominated by the various intrinsic defects originated from the off stoichiometry of the film composition. Moreover, deviation from the ideal stoichiometry during growth of this material might produce several secondary phases that includes, Zn(S,Se) Sn(S,Se), $Cu_2(S,Se)$, $Cu_2Sn(S,Se)_3$ etc. [10,11]. Therefore, to achieve the optimized material quality of the CZTSSe material, an extensive study of this material with various compositions is indispensable. In this paper, we have systematically varied the Cu/Sn CZTSSe absorber-layer to study the effect of compositional variation on the electrical, optical, and structural properties of the film.

2. Experimental

Polycrystalline CZTSSe thin films with typical thickness of 1.25 μm were gown over Mo coated sodalime glass (SLG) substrates using sulfo-selenization process. Samples were grown by Energy Solution Business Center, Showa Shell Sekiyu KK, Japan. Several CZTSSe samples with various Cu/Sn ratios in the film were prepared to study the effect of compositional variation, more specifically Cu-content on the films. Cu/Sn ratio in the films is determined based on the starting material precursor. S/Se ratio in all the samples is kept constant. Solar cell structure is consisted of *n*-ZnO/*i*-ZnO/CdS/$Cu_2ZnSn(S,Se)_4$ /Mo/SLG. The composition of the grown CZTSSe films was measured by electron probe micro-analysis (EPMA) at 15 kV of acceleration voltage. The structural properties of the films were examined by scanning electron microscopy (SEM). X-Ray diffraction (XRD) was taken in the range of 15°~ 90° by philips X'pert diffractometer at θ-2θ mode with Cu-Kα (α =1.541837 Å) radiation operated at 40 kV and 30 mA. Electrical properties were studied by the measuring the current-voltage (*I-V*) at dark using solar simulator. Optical band-gap was analyzed from the investigation of the external quantum efficiency (EQE) measurement of the solar cell structure. All characterizations were performed at room temperature.

3. Results and Discussion

3.1. Structural properties

For the study of the structural properties of the CZTSSe thin films, sample structure was considered as CZTSSe/Mo/SLG. Shown in figure 1(a)-(c) are the SEM images of the surface view of the several CZTSSe thin films grown with various Cu/Sn ratios in the films.

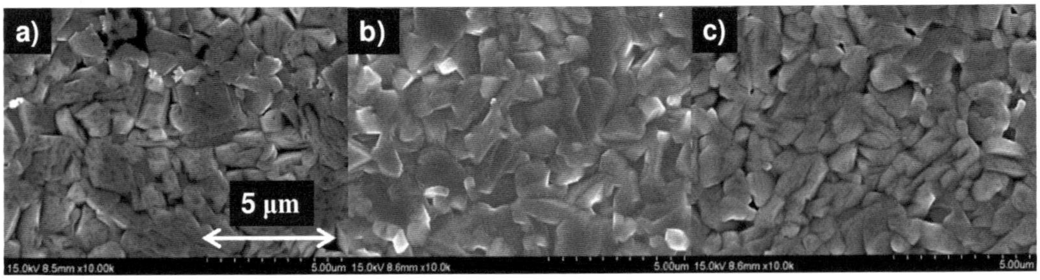

Figure 1. SEM images of the surface view of the several CZTSSe samples grown with a) Cu/Sn = 1.65, b) Cu/Sn = 1.75, and c) Cu/Sn = 1.85.

From the surface view, it is apparent that Cu-content in the films influences the morphology and the structure of the film including the grain size and shape. Film with Cu/Sn = 1.75 shows the morphology with large and uniform grains as seen from the Figure 1(b). No void or impurity phases were seen on the surface of the film. With decreasing the Cu-content in the film with Cu/Sn = 1.65 (Figure 1(a)), no apparent reduction in the grain size was observed. However, overall morphology deteriorates. There are some voids appeared inside grains. With a increase of the Cu-content (Cu/Sn = 1.85) grains become smaller in size with increasing number of voids inside grain boundaries. The shape also tends to be little granular for this sample. Smaller grains and voids strongly influence electrical properties of the polycrystalline material. Grain boundary act as scattering and recombination center for free carriers in the semiconductor. Thus increased grain-boundary increases the resistivity and reduces free carrier concentration of the material [12].

Shown in figure 2 is the XRD data for several CSTSSe films taken in the range of $2\theta = 10^{\circ}$ to 90°. XRD was taken on the structure of CIGS/Mo/SLG. No apparent differences have been observed in the XRD data for several samples with different Cu/Sn ratio in the film. The presence of (002) and (101) peak suggest that crystal structure of the films is either kesterite or stannite-type [13]. All the samples show well resolved (112) peak which is larger in intensities than the (220)/(204) peak. It indicates that irrespective of the growth condition, grains of the CSTSSe crystals find (112) orientation as kinetically preferable and grains continue to develop along this direction. Since the major diffraction peaks of CZTSSe overlaps with those of $Cu_2Sn(S,Se)_3$, $Cu_2(S,Se)$ and Zn(S,Se) phases, it is difficult to distinguish those phases from the CZTSSe phase in the XRD data. However, quantitative phase analysis of the rietveld refinement of XRD data of a separate CZTSSe sample (not shown here) reveals that significant amount of $Cu_2(S,Se)$ and Sn(S,Se) co exist with the CZTSSe phase.

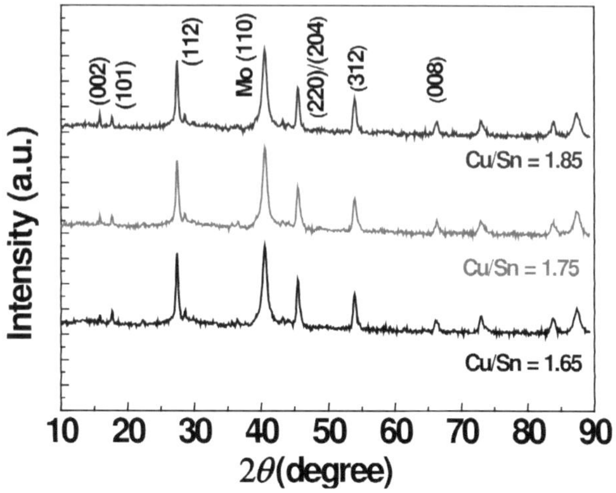

Figure 2. XRD data of several CZTSSe samples grown with a) Cu/Sn = 1.65, b) Cu/Sn = 1.75, and c) Cu/Sn = 1.85

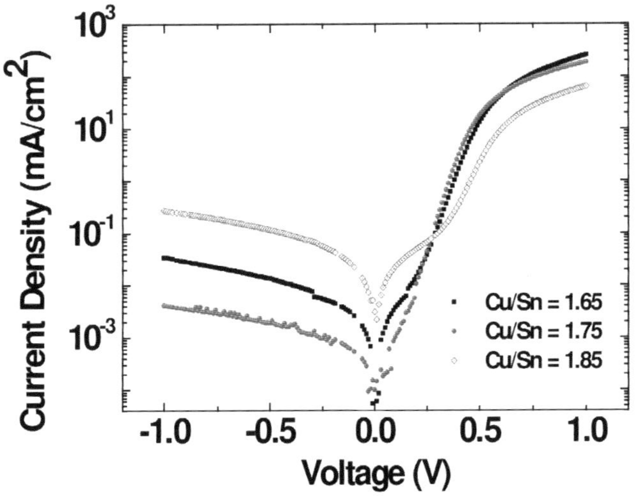

Figure 3. Dark *I-V* curve of the fabricated CZTSSe based solar cells where CZTSSe absorber-layers were grown with various Cu/Sn ratios in the film.

3.2. Electrical properties

Electrical properties of the CZTSSe samples were investigated by measuring the dark *I-V* curve of the solar cell structures. *I-V* curve of a *p-n* junction solar cell can be presented using single diode model as follows [14]:

$$J = J_0 \exp\left[\frac{q}{nkT}(V - RJ)\right] + G(V - RJ) - J_L \tag{1}$$

Here, J is the diode current density, J_0 denotes the saturation current density, q is the elemental charge, n is the diode ideality factor, k denotes the Boltzmann constant, T is the temperature, V is the applied voltage, R is the series resistance, G is the shunt conductance, and J_L is the photo current density. For the measurement at the dark, we can consider $J_L = 0$. In the forward bias region, the equation can be simplified as:

$$J = J_0 \exp\left[\frac{q}{nkT}(V - RJ)\right] \tag{2}$$

Then, ideality factor and saturation current density at a particular temperature can be determined from the slope of the Ln(J)-V curve. Figure 3 shows the dark *I-V* curve of all the solar cell structures fabricated with CZTSSe absorber layer. Flatten of the curves after V ~ 0.5 volt is due to effect of series resistance. It is apparent from the figure that sample with Cu/Sn = 1.75 exhibits lowest leakage current, while sample with Cu/Sn = 1.85 has the highest leakage current. Table-1 lists extracted diode ideality factor and saturation current density for all the samples. Series resistance was also calculated and listed in the table.

Table 4. Electrical parameters of the CZTSSe Solar cells extracted from the fitting of the dark *I-V* curve.

Cu/Sn	Ideality factor (n)	Saturation current density, J_o (mA/cm^2)	Series Resistance (Ω-cm^2)
1.65	1.64	1×10^{-4}	1.7
1.75	1.38	4×10^{-5}	2.2
1.85	2.25	4×10^{-4}	5.4

As seen from the table, all the samples show relatively higher ideality factor. It suggests that the performance of the solar cells is controlled by the recombination through localized states in the depletion region of the absorber layer. In particular, ideality factor of 2.25 together with higher leakage current of 4×10^{-4} mA/cm^2 and higher series resistance in case of the sample with Cu/Sn = 1.85 indicate that current transport properties of this sample is also dominated by recombination at the interface as well as recombination through trap assisted tunnelling. Higher series resistance also affect fill factor and current density of the solar cells.

3.3. Optical Properties

Optical band-gap of the CZTSSe materials has been determined form the measurement of the external quantum efficiency (*QE*) of the completed device. Assuming that all the carriers generated in the depletion region, *W* are collected without recombination loss and those generated within the diffusion length, *L* in the neutral region diffuse to the depletion edge, *QE* of the CZTSSe based solar cell can be approximated as [15]

$$QE\ (h\nu) = 1 - \frac{\exp[-\alpha(h\nu)W(V)]}{\alpha(h\nu)L+1} \tag{3}$$

Where, $\alpha(h\nu)$ is the absorption coefficient, $h\nu$ is the energy, and *V* is the bias voltage. At a particular bias voltage, assuming $\alpha(h\nu)L << 1$, the above equation can be more simplified as

$$QE\ (h\nu) \sim 1 - \exp[-\alpha(h\nu)W] \tag{4}$$

$$\alpha(h\nu) \sim -\ln(1 - QE(h\nu)) \tag{5}$$

Thus, QE in the sub-band gap region is proportional to the optical absorption coefficient of the material. Since, the optical band-gap is related to the photon energy near the band edge through the relationship [16],

$$\alpha h\nu \sim (h\nu - E_g)^{1/2} \tag{6}$$

Thus, it is possible to estimate the band-gap, E_g form the long wavelength tail of the *QE* plot against the incident photon energy, *E*. Shown in figure 4 is the optical band-gap of CZTSSe films grown with different Cu/Sn ratio in the films. From the figure it has been seen that optical band-gap of all the three samples is around 1.06 eV. A small difference is apparent among the band-gap of the films. sample with Cu/Sn = 1.85 shows relatively smaller band-gap. As mentioned earlier, it has been reported that chalcopyrite CZTSSe films contains several impurity phases which may include $Cu_2Sn(S,Se)_3$, $Cu_2(S,Se)$, $Zn(S,Se)$, $Sn(S,Se)_2$, $Cu_4Sn(S,Se)_4$ etc. Of course the existence of impurity phases depends on the growth condition and growth technique of the CZTSSe material. Nevertheless, electrical and optical properties of the CZTSSe material are determined by those impurity phases

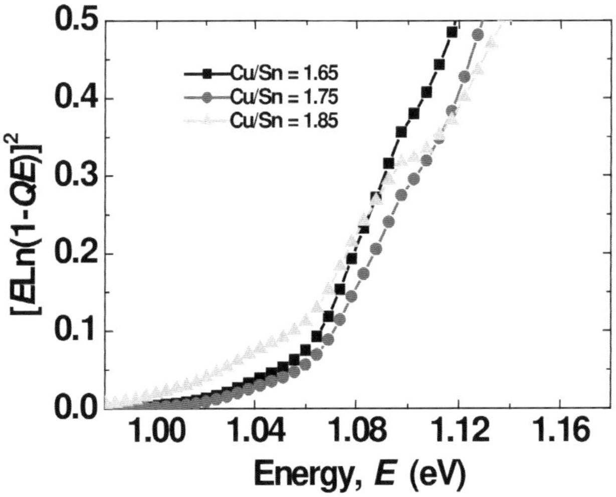

Figure 4. Optical band-gap energy of the CZTSSe thin films, determined from the long wavelength tail of the *QE* curve of the complete solar cell.

included in the material. For example, carrier concentration and optical band-gap of the CZTSSe material depend on the conduction-type and band-gap of the impurity phases. From the EPMA data it has been seen that for the samples with Cu/Sn ratio of 1.65, 1.75, and 1.85, Zn/Cu ratio are 0.58, 0.56, and 0.53 respectively. Therefore, Zn-rich phase is possible to be formed in the sample with Cu/Sn = 1.65, while Cu-rich phase may be formed in the sample with Cu/Sn = 1.85. Therefore, there might be some anomalies in the band-gap of the CZTSSe films arising from those impurity phases. Moreover, it can be seen from the figure 4 that there is band-gaps gradient in all the samples which suggests compositional inhomogeneities in all the CZTSSe samples.

4. Conclusions

CZTSSe thin films with various Cu/Sn ratio in the films have been investigated to study the effect of compositional variation over the electrical, optical, and structural properties of the film. It has been seen that surface morphology and grain size are significantly influenced by the Cu/Sn ratio in the films and morphology was found better for the samples with moderate Cu/Sn ratio of 1.75. Irrespective of the growth condition and compositional variation, all the CZTSSe crystals show (112) orientation as evident from the room temperature XRD data. In consistent with the SEM data, dark *I-V* curve reveals that that sample with Cu/Sn = 1.75 exhibits lowest leakage current, while sample with Cu/Sn = 1.85 has the highest leakage current along with larger ideality factor indicating larger recombination centers in this film. An anomaly in the optical band-gap has been explained with the presences of impurity phases and compositional inhomogeneities in the CZTSSe materials.

5. Acknowledgments

This work is performed under the program supported by the New Energy and Industrial Technology Development organization (NEDO) and the Ministry of Economy, Trade and Industry (METI), Japan.

References

[1] Tanaka T, Nagatomo T, Kawasaki D, Nishio M, Guo Q, Wakahara A, Yoshida A and Ogawa H 2005 *J. Phys. Chem. Sol.* **66** 1978
[2] Katagiri H, Jimbo K, Yamada S, Kamimura T, Maw W S, Fukano T, Ito T and Motohiro T 2008 *Appl. Phys. Express* **1** 041201
[3] Scragg J J, Berg D M and Dale P J 2010 *J. Electroanal. Chem.* **646** 52
[4] Chen S, Gong X G, Walsh A and Wei S –H 2009 *Appl. Phys. Lett.* **94** 041903
[5] Perrson C 2010 *J. Appl. Phys.* **107** 053710
[6] Shockley W, Queisser H 1961 *J. Appl. Phys.***32**, 510
[7] Wang W, Winkler M T, Gunawan O, Gokmen T, Todorov T K, Zhu Y and Mitzi D B 2013 *Adv. Energy Mater.* **4** 1301465
[8] Shin B, Gunawan O, Zhu Y, Bojarczuk N A, Jay Chey S and Guha S 2013 *Prog. Photovolt: Res. Appl.* **21** 72
[9] Jackson P, Hariskos D, Lotter E, Paetel S, Wuerz R, Menner R, Wischmann W and Powalla M 2011 *Prog. Photovolt: Res. Appl.***19** 894
[10] Dudchak I and Piskach L 2003 *J. Alloys Compd.* **351** 145
[11] Nagoya A, Asahi R, Wahl R and Kresse G 2010 *Phys. Rev. B* **81** 113202
[12] Kazmerski L L, Berry W B and Allen C W 1972 *J. Appl. Phys.***43** 3515
[13] Xianzhong Lin, Jaison Kavalakkatt, Kai Kornhuber, Daniel Abou-Ras, Susan Schorr, Martha Ch. Lux-Steiner and Ahmed Ennaoui 2012 *RSC Advances* **2** 9894
[14] Sze S M 1969 *Physics of Semiconductor Devices* (New York: Wiley–Interscience)
[15] Liu X X and Sites J R 1994 *J. Appl. Phys.* **75** 577
[16] Pankove Jacques I 1971 *Optical Process in Semiconductors* (New York: Dover)

AUTHOR INDEX

Institute of Physics
Dirac House, Temple Back
Bristol BS1 6BE UK

proceedings.com